T0347301

THE FUTURE OF OIL
AS A SOURCE OF ENERGY

THE FUTURE OF OIL
AS A SOURCE
OF ENERGY

THE EMIRATES CENTER FOR STRATEGIC
STUDIES AND RESEARCH

THE EMIRATES CENTER FOR STRATEGIC STUDIES AND RESEARCH

The Emirates Center for Strategic Studies and Research (ECSSR) is an independent research institution dedicated to the promotion of professional studies and educational excellence in the UAE, the Gulf and the Arab world. Since its establishment in Abu Dhabi in 1994, ECSSR has served as a focal point for scholarship on political, economic and social matters. Indeed, ECSSR is at the forefront of analysis and commentary on Arab affairs.

The Center provides a forum for the scholarly exchange of ideas by hosting conferences and symposia, organizing workshops, sponsoring a lecture series and publishing original and translated books and research papers. ECSSR also has an active fellowship and grant program for the writing of scholarly books and for the translation into Arabic of works relevant to the Center's mission. Moreover, ECSSR has a large library including rare and specialized holdings and a state-of-the-art technology center, which has developed an award-winning website that is a unique and comprehensive source of information on the Gulf.

Through these and other activities, ECSSR aspires to engage in mutually beneficial professional endeavors with comparable institutions worldwide, and to contribute to the general educational and scientific development of the UAE.

First published in 2003 by
The Emirates Center for Strategic Studies and Research
PO Box 4567, Abu Dhabi, United Arab Emirates
Website: http://www.ecssr.ac.ae
E-mail: root@ecssr.ac.ae

Copyright © The Emirates Center for Strategic Studies and Research, 2003

Distributed by RoutledgeCurzon
11 New Fetter Lane, London EC4P 4EE

British Library Cataloguing in Publication Data
A catalogue record for this book is available from the British Library

Library of Congress Cataloguing in Publication Data
A catalogue record for this book is available from the Library of Congress

ISBN 9948-00-010-2 hardback
ISBN 9948-00-009-9 paperback

Contents

Contents

The Oil Industry: Issues of Concern

FIGURES AND TABLES

FIGURES

TABLES

ABBREVIATIONS AND ACRONYMS

AAPG	American Association of Petroleum Geologists
ABB	Asea Brown Boveri
AFC	alkaline fuel cell
ANZ	Australia and New Zealand
API	American Petroleum Institute
ARCO	Atlantic Richfield Company
AWEA	American Wind Energy Association
bbl	blue barrels
BCM	billion cubic meters
BCR	benefit–cost ratio
BMW	Bayerische Motoren Werke
boe	barrels of oil equivalent
BP	British Petroleum
BTU	British Thermal Unit
CARB	California Air Resources Board
CBM	coal bed methane
CCGT	combined cycle gas turbine
CDIAC	Carbon Dioxide Information Analysis Center
CDM	clean development mechanism
CEED	Center for Energy and Economic Development
CET	clean energy technologies
cf	cubic feet
CIF	cost, insurance and freight
CO_2	carbon dioxide
cp	centipoise
CP	cumulative production
CSEG	Canadian Society of Exploration Geophysicists
CT	compliant tower
DHFCV	direct hydrogen fuel cell vehicle
DOE	(US) Department of Energy
DOI	(US) Department of Industry
DTI	Department of Trade and Industry
E/GDP	energy/real gross domestic product
EIA	Energy Information Administration
EPA	(US) Environmental Protection Agency

EPRI	Electricity Policy Research Institute
ERG	Energy and Resources Group
EROI	energy return on investment
EU	European Union
EV	electric vehicle
FCV	fuel cell vehicle
FP	fixed platform
FPS	floating production systems
FSU	Former Soviet Union
FT	*Financial Times*
Gb	gigabarrels (billion barrels)
GCC	Gulf Co-operation Council
GDP	gross domestic product
GEF	Global Environment Facility
GHG	greenhouse gases
GM	General Motors
GNP	gross national product
GtC	giga (billion) tons of carbon
GTL	gas-to-liquids
IEE	Institution of Electrical Engineers
ICE	internal combustion engines
IEA	International Energy Agency
IEO	International Energy Outlook
IIASA	International Institute for Applied Systems and Analysis
IMF	International Monetary Fund
IPCC	Intergovernmental Panel on Climate Change
IT	information technology
J	Joule
JIP	joint industry project
JNOC TRC	Japanese National Oil Company Technology Research Center
kWh	kilowatt hour
LDCs	less developed countries
LEV	low emission vehicle
LNG	liquefied natural gas
MB	megabytes
mb/d/yr	million barrels per day per year
MCFC	molten carbonate fuel cell
MIT	Massachusetts Institute of Technology

MMS	Mineral Management Services
MTOE	million tons of oil equivalent
MTP	market transformation program
MWD	measurement while drilling
NASA	National Aeronautics and Space Administration
NEB	National Energy Board
NGL	natural gas liquids
NPD	Norwegian Petroleum Directorate
NPV	net present value
NREL	National Renewable Energy Laboratory
NZ	New Zealand
OAPEC	Organization of Arab Petroleum Exporting Countries
OECD	Organization for Economic Cooperation and Development
O&GJ	*Oil and Gas Journal*
O&M	operation and maintenance
OMI	Oil Market Intelligence
OPEC	Organization of Petroleum Exporting Countries
ORNL	Oak Ridge National Laboratory
PAFC	phosphoric acid fuel cell
PCAST	President's Committee of Advisors on Science and Technology
Pcf	Peta cubic feet
PEM	proton exchange membrane
PIW	*Petroleum Intelligence Weekly*
PM	particulate matter
ppmv	parts per million volume
PR	progress ratio
PS	petroleum system
P&T	pressure and temperature
PVs	photovoltaics
RAEL	Renewable and Appropriate Energy Laboratory
rbl	red barrels
R&D	research and development
REPP	Renewable Energy Policy Project
ROV	remote operating vehicle
ROW	rest of the world
R/P	Reserves/Production
RR	remaining reserves
RRI	Robertson Research International

SCSSV	sub surface safety valve
SEC	Securities and Exchange Commission
SEI	Stockholm Environment Institute
SI	International System of Units
SOFC	solid oxide fuel cell
SPE	Society of Petroleum Engineers
SPFC	solid polymer fuel cell
SS	subsea system
STEO	short term energy outlook
SUV	Sport and Utility Vehicle
tb/d	thousand barrels per day
Tcf	tera cubic feet (trillion cubic feet)
TES	Transportation Energy Strategy
TLP	tension leg platform
toe	tons of oil equivalent
TVD	total vertical depth
UAE	United Arab Emirates
USDOE	US Department of Energy
USDOI	US Department of Industry
USEPA	US Environmental Protection Agency
USGS	US Geological Survey
VLEV	very low emission vehicle
W	Watt
WD	water depth
WEC	World Energy Council
WO	*World Oil*
Wp	Watts-peak
WPC	World Petroleum Congress
WTI	West Texas Intermediate
ZEV	zero emissions vehicle

FOREWORD

Oil has always been regarded as a key component of the crucial energy sector, which drives economic progress. However, the global predominance of oil as a source of energy has been affected in recent years by several developments. Positive and negative determinants are shaping the future prospects of oil. Given the finite nature of the resource, the relative merits of alternative sources of energy, the possibility of major technological breakthroughs and new environmental concerns and imperatives, significant shifts are likely to take place in the global energy scenario.

Greater diversification in energy sources and the growing use of alternative and renewable forms of energy have been prompted by economic, ecological and security considerations. Non-conventional sources, such as photovoltaics, fuel cells, microturbines and wind power, are becoming increasingly viable and pose challenges to more conventional sources of energy. All indications are that the fossil fuel era will not last indefinitely and that the world is undergoing a transition from a hydrocarbon-based economy to one based on renewable forms of energy.

The future of oil is a matter of profound concern not only to petroleum companies, but also to oil-producing nations and oil-dependent economies. Energy experts who assembled in Abu Dhabi for the ECSSR Sixth Annual Energy Conference, held on October 7–8, 2000, focused attention on various aspects relating to *The Future of Oil as a Source of Energy*. The main thrust of their presentations, which are compiled in this book, identifies the significant trends and factors influencing the global position and price of oil.

Among the important topics discussed at the conference were the realities of oil as a finite resource, the complexities of energy forecasting and the positive and negative drivers affecting the position of oil. The comparative advantages and disadvantages of different energy sources were highlighted, and the viability of alternative and renewable sources assessed. Furthermore, the rising importance of natural gas was analyzed in the context of the trend toward decarbonization and "clean energy" sources. The need to control volatility and to maintain stability in the oil market was underlined.

In the ultimate analysis, it would appear that resource availability is less likely to determine the future of oil than other economic, ecological and technological considerations. Collectively, the conference presentations

[xvii]

afford a clearer picture of the current energy transition and the anticipated position of oil in the emerging scenario. For the oil producers and oil-dependent economies of the Arabian Gulf, valuable guidelines are offered on how to meet the inevitable challenges posed by an era in which oil may no longer be the prime source of energy.

ECSSR takes this opportunity to thank all the distinguished participants of the Sixth Annual Energy Conference for their stimulating presentations. We would also like to express our appreciation to the panel of expert referees who reviewed these papers and offered their valuable comments. ECSSR is grateful to Mr. Robert Mabro CBE, Director of the Oxford Institute for Energy Studies, for sharing his expert views on the subject through his introductory chapter and concluding observations. A word of thanks is also extended to ECSSR editor Mary Abraham for coordinating the publication of this book.

Jamal S. Al-Suwaidi, Ph.D.
Director
ECSSR

Section 1

INTRODUCTION

The Future of Oil

Robert Mabro

D oes oil have a future? This question is being asked from time to time by those who live in countries heavily dependent for their economic well-being on revenues accruing from petroleum exports. I also suspect that oil companies concerned about business prospects in the very long run ask themselves this question. BP's recent choice of the "Beyond Petroleum" motto may suggest that this concern is genuine, although some cynical commentators believe that the only purpose is to placate environmentalists and other petrophobic groups.

The question, no doubt, is legitimate. Oil, like gas or coal, is a physical resource. Nature may have been very generous but the size of its mineral gifts is finite. Sooner or later, all the reservoirs that exist on our planet will be discovered. And, if the demand for fossil fuels remains positive for as long as supplies are forthcoming, full depletion will inevitably occur. On this assumption, supplies will dry up at some future date, which is as yet highly uncertain.

However, physical exhaustibility is not the only threat to oil, or for that matter to any commodity, good or service. Primary commodities, whether exhaustible or renewable, often give way to substitutes, which in turn give way to other commodities that emerge on the world scene. In energy, firewood gave way to coal, then coal to oil. In other fields, plastics replaced metals in a wide variety of uses; artificial fibers superseded cotton, silk and wool. In all cases, however, the original commodity always continues to be used, retaining a small share of demand, or carving for itself a niche in the market. Thus, firewood is still being used as a fuel, not only in developing countries; coal still supplies some 20 per cent of world energy; natural rubber remains irreplaceable in the manufacturing of aircraft tires; and silk and cotton are sought after by high-income customers who can afford to dislike the airtight texture of artificial fabrics.

Primary products are not the only victims of substitution. There is a product cycle for manufactured goods. The invention of color TV eventually kills the black and white set, which for a time retains a share of the demand from low-income groups until the color TV becomes affordable by most consumers. The typewriter has given way to the computer and the slide-rule to the hand calculator. One can multiply the examples *ad infinitum*.

However, the fact that a displaced commodity or product retains some use for a long time is of no great consolation. Those who worry about the future of oil are concerned about the size and the growth of its market. To know that some oil will be consumed for a long time in some small markets or for some esoteric purposes does not alleviate these worries.

Thus the issue remains whether oil is threatened by an economic demise that may precede, perhaps by a long time, the physical exhaustion of the natural resource. The threat to oil comes from two sides: on the supply as well as the demand fronts.

Let us first consider the supply aspect. It is important to recall in this context that crude oil is not consumed as such, except in very rare situations. Crude oil is an intermediate good. The final demand is overwhelmingly for petroleum products obtained from refining the raw material. Some of these products have substitutes. Fuel oil, for example, which is used to raise heat under boilers, can, and indeed has, been replaced by primitive biomass, coal, natural gas or sophisticated nuclear. Gas oil, when used for space heating, competes with natural gas, coal, wood and primary electricity. Other petroleum products are, however, at the other end of the substitution spectrum; there are no readily available substitutes. These are the automotive fuels – gasoline and diesel – and the fuels for aircraft – jet kerosene and aircraft gasoline. Products used as petrochemical feedstock have imperfect substitutes.

The oil domain, therefore, is the transport sector. The threat, on the supply front, in this sector can come from the discovery of a new fuel that can power cars and airplanes both economically and efficiently. Research is under way in the search for a new product that meets these conditions. It should be recalled here that a change in fuel requires changes in the engines that will use it. For batteries to replace gasoline, one needs to develop new types of batteries and design the electric car. It seems that the latter is easier than the former.

The fuel cell, a technology that elicits much enthusiasm in some quarters and much skepticism in others, will also be associated with some adaptations, minor as they may be, to the engine. But this is a small side issue. The important ones relate to the likely time horizons – the time required for research to lead to discovery, for the discovery to be developed into an economic (that is cost-competitive) product, and for the building of the infrastructure required to introduce and market this product. A new fuel requires the adaptation of service stations for storage and sales; a new type of car will not replace the existing stock immediately since technical and economic obsolescence take time to set in. Another important question about new fuels relates to the energy required for their production. Electric batteries have to be charged, hydrogen has to be obtained. If the primary source is a hydrocarbon, then the future of oil may be less bleak than assumed.

The threat for oil on the demand side is said to come from environmental concerns. Climate change, which may be partially caused by the burning of fossil fuels, is inducing a wave of petrophobia in many countries. The debate over the Kyoto Protocol is familiar. The questions are, first, whether the international community will marshal the political will to cooperate in implementing measures that reduce effectively the volume of greenhouse gases and, second, whether these measures, implemented as agreed in Kyoto, will by themselves cause a significant reduction in the consumption of oil?

The reluctance of the US to agree with the Kyoto measures provides ammunition to the skeptical observers who believe that the agreement will not be implemented in a meaningful way. We shall have to wait and see whether the US will have a change of heart or not and, as importantly, whether a way will be found to involve the developing countries, particularly China and India, in a program which will inevitably fail if it does not obtain global adhesion.

Let us assume, however, that the Kyoto Protocol is implemented and that the cooperation of China and India is secured in a second phase. Yet, it does not follow from this assumption that such an implementation will result in a significant reduction in oil consumption as compared with a non-Kyoto, business-as-usual situation. The reasons are manifold. First, if carbon emissions are the issue, then coal should suffer more than oil and gas. Second, oil is entrenched in the transport sector where it cannot be easily replaced. Reductions in fossil fuel use will thus fall more heavily where they can be more easily substituted. Third, if permit trading is widely used, the impact on all fuels (and this includes oil) will be significantly mitigated. Fourth, industrialized countries will increasingly seek solutions involving carbon sinks to enable them to meet the targets without disturbing living patterns too much. Fifth, the Kyoto program is not very ambitious anyway. Countries are given the opportunity to avoid reducing emissions by large amounts as they can buy permits from Russia, which enjoys considerable credits. The reason is that Russia is emitting much less than in 1990 (the base year) because of the destruction of its heavy industry, which collapsed with the communist regime.

Some countries, particularly in Europe, may attempt to do more than is required by the Kyoto program, to curb energy consumption. The European Union's view on energy security is that dependence on imports should be curbed through demand management measures. As governments prefer revenue-raising measures to projects that involve increases in public expenditure, they are inclined to use fiscal instruments to manage demand. The easy solution is to tax automotive fuels heavily. This easy solution has been already applied to such an extent in Europe that further increases in petrol and diesel taxes have become politically unacceptable. If governments were serious about the need to throttle the growth in fuel consumption in transport, they would invest in public transportation systems and subsidize research and development (R&D) in car technology. However, they are reluctant to do so, as this means increases in public expenditures.

The question is whether the driving force in energy policy is security or the environment? These two objectives do not conflict if the policy aims at reducing the use of fossil fuels. However, they do conflict if it seeks lower dependence on oil and gas imports by promoting the use of coal or nuclear energy. Coal emits more CO_2 than oil or gas and is far from being environmentally friendly if the concern is global warming. Nuclear energy does not emit CO_2 but is profoundly disliked by environmentalists because of the radiation risks. Both the EU and the USA are talking again about the nuclear option, which was considered to be socially unacceptable until very recently. Nuclear energy poses a threat to fuel oil to the extent of its current use in power generation. Fuel oil, however, is being increasingly displaced by natural gas, and the growth potential of gas is largely in that sector. In other words, nuclear power represents a greater threat to gas than oil.

Some believe that the solution to both the environment and the security problem will be provided by renewable energy, particularly solar and wind. Renewables can be produced *in situ*: both solar and wind in many regions of the world; solar or wind in others. This reduces import dependence. Renewables do not emit CO_2, sulfur or particles. In this sense, they are environmentally friendly. Solar energy, however, is very expensive except where it is used to produce low-temperature heat. High-temperature heat is needed for power generation and recourse to solar energy in this sector involves very high costs. Windmills are noisy and ugly and as such environmentally objectionable. The solution is to build them offshore, which increases their costs.

Solar and wind energy also threaten fossil fuels in their use for power generation and for applications such as space heating/cooling, where secondary electricity can displace oil, gas and coal. Their impact on the transport sector depends critically on the development of electric cars, since the batteries that will power them have to be charged. We are still very far from an era in which energy supplies will be dominated by the sun and wind, and probably still far (even if not as much) from the domination of nuclear energy.

We thus have two issues: depletion of finite reserves and economic demise caused either by substitutes, changes in the technology of oil-using vehicles or appliances or by policies that curb the demand for oil. These policies may be pursued for either environmental or security objectives or both.

Exporting countries endowed with large oil and/or gas resources need not worry very much about the issue of physical exhaustibility. The debate on whether the world will be running out of oil in the next 20 or 30 years, or whether the resources are so plentiful as to extend the depletion time horizon well into the second half of this century, has been recently revised by the works of C.J. Campbell and J.M. Laherrère. Their main study, *The World's Oil Supply 1930–2050*, was published in 1995 by PetroConsultants. Although the authors' views have since evolved toward a slightly less pessimistic outlook, their main arguments remain essentially the same as expounded in this major study. Their conclusions can be summarized as follows:

The phase of oil history in which we find ourselves now is the phase of transition between a period of growth (1959–70) and one of decline which is expected to set in at the beginning of [this] century ... Resource constraints will begin to be felt in most parts of the world save in the Gulf. By 2050 production would have fallen to the level of 1950, that is some 18 million barrels/day ... Of course, some oil will be left after 2050, an estimate of 236 Gb.[1]

Critics have pointed out that these views are far too pessimistic because they do not give due weight to the following factors:

- the existence of huge reserves of non-conventional oil that can be profitably exploited if oil prices are sustained at the $25 per barrel level
- the advances in technology that increase dramatically the recovery factor from oilfields.

Insofar as oil-exporting countries with large reserves are concerned, the debate is irrelevant. Even if the threat of physical exhaustibility is relatively close, they will not be the first ones to be adversely affected. In any case, we do not think that this threat is imminent.

A more serious cause of worry is the competition they may suffer from the possible development of unconventional oil in Canada and Venezuela. Such a development will only happen if the energy security objective leads the USA to promote and subsidize it in order to reduce reliance on imports from the Middle East and realize the idea of Western hemisphere self-sufficiency, or if private investors become convinced that oil prices will be sustained (and sustainability is what really matters) at levels that make exploitation of tar sands, shale oil and the Orinoco Belt in Venezuela commercially viable. In other words, there is a threat to the future of Middle East oil, which does not arise from exhaustion but from other types of oil.

The economic demise of oil is a long-run issue because of the time lags involved in the development of new technologies and the penetration of markets. There is no doubt, however, that substitutes will eventually emerge and that the transport technology will be revolutionized. As regards substitutes, gas will continue to increase its share of the energy consumption mix, particularly in the power generation sector. If the environmental concern overrides the security objective in industrialized countries, it will displace coal. Otherwise, coal will receive a boost in the USA and Europe even though it may give way to both oil and gas in Asia. Nuclear energy may revive if the security issue becomes paramount in the West; and in any case it is likely to increase its share in Asia, where the expansion of gas use may hit constraints due to either politics or high transport costs or both. Nuclear energy has a better chance than solar and wind energy, provided that its acceptability is not undermined further by some serious accident.

As regards transport technology, we shall have to wait and see. Sometimes technological breakthroughs expected with some trepidation do not materialize and sometimes they occur suddenly, taking the skeptics by surprise. Looking ahead, we can consider the future as consisting of two successive, albeit overlapping, periods. The first may take us at the earliest to 2020 and perhaps as late as 2030.

The problem facing the oil world in that period will not relate to reserves but to the development of production capacity. At times, there will be too much capacity, causing potential gluts and depressing prices, and at times too little capacity, causing price spikes with their inevitable adverse consequences for the world economy and the subsequent levels of oil demand. Capacity cycles arise for either political or financial reasons. Managing capacity is the real challenge, and neither OPEC nor the private companies have addressed it as yet. As mentioned earlier, the competition between conventional and non-conventional oil is likely to become an important issue as we advance in that period.

In the second period, which may begin around 2020, we are likely to hit a peak in the production of conventional oil, which will induce greater investments in non-conventional oil, natural gas and perhaps nuclear energy. The initial inducement may come from a price spike. But even if the price rise is not sustained, the perception that a production peak has been reached will cause expectations of future price rises. This transition may not be very smooth, since the emergence of new supplies induced by price expectations may instead cause prices to fall. As usual, the fear of a shortage leads to the emergence of a glut, which is perhaps temporary. Substitutes and new car technologies will appear and begin to penetrate the market in this second period and slowly threaten the share of oil in energy consumption.

However, one should recall that another force will be operating relentlessly in times of economic growth throughout these periods. This is energy demand growth. The demand for all fuels will expand in the good years by

2–3 percent per annum. The demand for oil will also continue to grow because consumption patterns in developing countries will increasingly include more cars, more modern forms of space heating, more air travel, more electricity. The developing world is still very far from the levels of per capita energy use in industrialized countries and, although catching up will take several decades, much more energy, including much more oil, will be demanded in the years to come. We can therefore look at a period ahead of us during which the share of oil in the energy mix may well shrink but the amounts produced and consumed will continue to increase. In the final analysis, absolute volumes count for more than relative shares.

All that should not encourage any complacency. Countries that depend heavily on oil revenues face the daunting challenge of transforming their economies to enable them to grow in an era of reduced oil income. The time available for this transformation is of the order of 20 or 30 years, but this is not as long as may appear at first sight. In the early 1970s many among us thought that the newly accrued oil wealth should enable the Gulf countries to build the foundations of a non-oil economy capable of sustaining long-term economic growth. Thirty years have elapsed and these foundations are not yet there.

To prepare for an economic future in the period when the prospects for oil will begin to decline requires immediate action. The task should begin today. The policies that favor future economic development do not bear fruit very quickly. Their implementation takes time. They only yield the desired results after very long gestation periods. To delay the formulation and execution of these policies will put the economic well-being of the oil nations in serious jeopardy. The required policies apply to three different areas.

The first of these areas is investment in human capital. This calls for a radical reform of the education system in the oil-exporting countries concerned. At present, education develops the memory more than analytical abilities. It values authority more than the small but necessary amount of freedom that encourages the creative mind. To define and introduce a new educational philosophy is an immensely difficult task, but it is not impossible. The process will inevitably be slow, and this is why it needs to be initiated today.

The second area is the labor market, in which the current pattern of incentives does not favor work in the productive sectors. Even the oil industry, the source of current wealth for exporting countries, fails to attract all the talent needed for its development. A reform of labor laws, which currently cause private employers to prefer migrant workers to nationals, should also be considered. The present situation involves a serious distortion characterized by rising unemployment among nationals and increasing reliance on foreign migrants.

The third area is regional cooperation. All the Gulf countries are too small in size. Even the larger oil-exporting nations such as Saudi Arabia would benefit from an expansion of their markets. At present, despite the work done by the Gulf Co-operation Council (GCC), the degree of economic cooperation in the Gulf region still leaves something to be desired. Cooperation will enlarge the size of the market and thus encourage investment in industries benefiting from economies of scale and from a potentially significant volume of demand. Cooperation will reduce the incidence of wasteful duplication of projects and activities, and will increase the efficiency of both labor and capital.

Whether the future of oil is threatened or assured, the wise course of action is to prepare for a day when revenues may not be sufficient to maintain living standards. If oil turns out to be under threat, the development of a non-oil economy will provide a safety net. If the optimists turn out to be correct in their view that oil will continue to have a future for a very long time, the development of a non-oil economy will add to the wealth of countries. Nothing will be lost in either case.

POSITION OF OIL: AN OVERVIEW

1

Oil as a Source of Energy:
Present Realities and Future Prospects

Jean Laherrère

Forecasting is a difficult task, even if it is based on a huge amount of valid data, as demonstrated by weather forecasting. However, in the case of oil, forecasting is infinitely more difficult because the data on which it is based is often uncertain and questionable.

Remaining reserves, which indicate future production, are estimated by using geological, geophysical and engineering data from wells and seismic surveys. Furthermore, decline analysis can give a good indication once a field is in decline. In practice, however, the reporting of oil data is often influenced by political considerations and depends on the desired image to be projected.

For example, technical estimates of reserves, which are normally given as a range of values, often differ from reported reserves, for which a single value is quoted. Maximum estimates can be used to secure higher OPEC quotas or to promote share values or share options, whereas minimum estimates can reduce taxes and royalties, and allow progressive upward revisions when production exceeds the estimate, which help to balance assets over periods of lean discovery.

The units and symbols used by the oil industry are likewise archaic and unregulated. In addition, definitions are often imprecise, with various meanings attributed to the same term by different authors. Ambiguity is preferred because it provides an opportunity for the author to report what suits him best.

It is very hard to forecast oil prices because they depend on consumer behavior and political decisions. Economic and technological developments shape future oil supply and demand. The perception of future shortage or abundance also influences the situation, affecting the important derivatives market in particular.

Imprecise definitions, poorly reported data and genuine geological and physical uncertainties generally result in poor forecasting. This is a case of GIGO: garbage in, garbage out. Estimates of the quantity of undiscovered oil and reserves range from 200 Gb to 1700 Gb.

Such a broad range has little practical value. One needs to plan for the future on the best realistic estimate, not on fantasies. Therefore, the challenge is to secure better data, rather than find a better method of evaluation, for existing methods are already quite adequate for the task. It would be very much in the interest of the national companies, especially in the Middle East, if they were to publish more accurate information, letting the world know what their indispensable role truly is.

The Definition of Oil

Oil can represent many different substances, including crude oil as such and crude oil combined with one or more of a number of substances, namely lease condensate, natural gas liquids, synthetic oil and liquids. Lease condensate, as measured at the wellhead separator, is present during the gas phase under reservoir conditions, but becomes liquid at surface temperature and pressure. In the United States crude oil and lease condensate are measured together. Natural gas liquids (or NGL) are processed from gas in a plant but, confusingly, condensate is sometimes included as NGL. Synthetic oil is derived from upgrading heavy oil, tar sands or shale oil (as planned in Australia), or from the conversion of coal or gas by the Fischer-Tropsch process. Liquids represent processing gains from refining oil.

Oil is divided into conventional and non-conventional or unconventional categories, but there is no consensus on where to draw the boundary between these. For some authors, conventional oil refers to the proceeds of primary and secondary recovery using only water or gas flooding, but excluding the proceeds from enhanced oil recovery, heavy oil (variously defined as below 20, 17.5, 15 or 10 degrees API), tar sands (defined by a high viscosity over

10 000 cp), oil shales (mainly immature source-rocks) and oil in difficult locations such as the Arctic and deep water. Others, such as the US Geological Survey, prefer to treat as conventional all oils with a well-defined down-dip water contact, which are significantly affected by the buoyancy of petroleum in water. They confine unconventional oil to so-called continuous-type deposits. This geological definition ignores factors such as water depth, regulatory status or the status of technology. In still other cases, conventional oil merely means cheap oil.

Lack of Standard Terminology

Every country in the world, except Liberia and Bangladesh, has accepted the International System of Units (SI), which is also known as the metric system. It is the mandatory official system in the European Union, Canada, Australia and, since 1993, even for US federal agencies.

The SI unit for volume is the cubic meter (m^3); for weight, the tonne (t); for energy as work and heat, the joule (J), as calorie and thermie are obsolete; for power, the watt (W) {1 W = 1 J/s; 1 kWh = 3.6 MJ}. A barrel (of oil) is not an official unit in the USA, which explains why federal agencies are obliged to define it as 42 US gallons.

However, the industry has been very slow to adopt the metric system and persists with confusing, archaic and cumbersome units, such as M for thousand (k), MM for million (M) and MMM or B for billion (G). Failure to pay proper attention to units can have serious consequences, as is illustrated by the loss of the Mars Climate Orbiter. This incident happened because NASA sent the instructions for thrust in newton (force which accelerates 1 kg at 1 m/s^2), while Lockheed, the contractor, had designed it to operate according to pounds.

There is confusion about the use of symbols both by the industry and the media, although the SI system is unequivocal, providing k (kilo) for thousand, M (mega) for million, G (giga) for billion and T (tera) for trillion. These symbols are accepted in the computer business (MB = megabytes, GB = gigabytes) and for such things as Y2K. Even the symbol for barrel is variously quoted as B, b, bl and bbl. The latter incidentally stands for blue barrel, because in earlier years crude oil was sold in blue barrels (bbl) and refined products in red barrels (rbl). Production is given in BPD or BD, instead of b/d.

[17]

Equivalence is often used to compare energy in terms of barrels of oil equivalent (boe). However, that too is confusing because there are many different kinds of crude oil and many different categories of equivalence. Calorific equivalence gives 1 boe = 5.6 or 6 kcf at the wellhead, but, since gas costs five to ten times more than oil to transport, an equivalence in value is about 1 boe = 10 kcf. Transport of NGL consumes about 12 percent of the energy content of the liquefied gas in the tanker. Gas-to-liquids (GTL) conversion is about 1 boe = 10 kcf. Canada uses price equivalence for each year at the burner tip, where 1 boe equalled 14 kcf in 1989 and 20 kcf in 1991.

The lack of precise standard definitions and terminology is the cause of much confusion and unnecessary argument. "Gas" is a colloquial term for gasoline in the United States that leads to confusion with natural gas. The term "reserves" means an estimate of the remaining amount of oil to be recovered (produced) from known fields. The term "resource" means the estimated amount in the ground. Only a small part of the resource will in fact be recovered. Reserves are often confused with resources. The term "recoverable reserves" is tautological, because by definition reserves are recoverable.

The term "reserves" is often used without stating what exactly is meant by it. It may mean remaining reserves at the reference date (this date must be indicated), or it may mean total recovery, which corresponds with cumulative production when the field is abandoned. The term "original (or initial) reserves" may be synonymous with total recovery or simply the first estimate. Reserves are variously termed "proved" in the US and "proven" in the UK.

Oil is not produced in the sense that wheat, corn or fruit are produced but is extracted, being pumped when it does not rise to the surface automatically. Oil is produced pure or together with water, gas or even sand. It is a liquid, not an ore, and is completely different from other mineral deposits such as coal, gold and copper, where extraction depends heavily upon concentration. The oil–water contact in an oilfield is abrupt, leaving no scope to tap lower concentrations, although the percentage of water produced with the oil rises toward the end of a field's life.

Volume and Weight

Put in simple terms, oil remains oil for the consumer, who is not interested in the many different sources from which it comes. Quoted oil-supply figures commonly indicate undifferentiated crude oil and condensate. Natural gas liquids are produced by removing the heavier hydrocarbons from gas production. In the Frigg gasfield of the North Sea, for example, the NGL represent no more than $1g/m^3$, but they are removed by a large plant for the sake of English consumers. The NGL yield from gas is 25 percent in the USA, compared to only 5 percent in the rest of the world. Condensate and NGL are commonly confused in the statistics. The Norwegian Petroleum Directorate (NPD) in their 1999 report give data as oil ($M.m^3$), gas ($G.m^3$), condensate ($M.m^3$) in volume, but NGL in weight (Mt with $1 t = 1.3 m^3$ oe). In the past only NGL values were reported. In the United Kingdom, the Department of Trade and Industry (DTI) gives oil production in tons (Mt), without distinguishing condensate and NGL, meaning that the gravity has to be known to convert the value to volume (average UK $1.199 m^3 = 1 t$). Gravity differs from field to field and even over time.

Ways to Deal with Uncertainty

Estimates of the potential reserves in a potential field are based on regional information concerning reservoir characteristics and seismic data, which give the structural geometry of the field with a resolution of within a few tens of meters. The estimates are progressively improved with data from the initial borehole (wildcat) and any delineation wells, so that the various parameters, such as net pay, porosity and oil saturation, may be measured more accurately.

The estimates are given as a range from minimum to maximum, together with the intervening values for the most likely estimate (or mode) and the mean estimate (weighted average). The uncertainty is generally described in terms of its converse, namely probability, with the minimum being 95 percent or 90 percent, the maximum being 5 percent or 10 percent, the mode around 65 percent in a lognormal distribution and the mean around 40 percent.

The probability of the estimate of the reserves for a field is in fact a subjective probability, as each case is unique and there is no means of verifying the chosen quantities save by a statistical analysis of a large number of cases. The very concept of probability is both hard to grasp and difficult to apply in practice. Many analysts prefer to stay with the so-called deterministic approach, relying on a single "best estimate," as they either do not understand probability theory or are not convinced of its applicability to estimate reserves. Figure 1.1 gives a lognormal distribution of the variation in size of reported reserves versus the probability of their being higher than indicated. The 50 percent probability value (P50 or median) is often used but is difficult to apply in practice as it is supposed to be the average of many unknown solutions. The concepts of minimum, maximum and most likely (mode) or best estimate are the only ones that the human mind can readily understand. The average may be calculated as a third of the sum of mini + mode + maxi values.

Figure 1.1
Distribution Probability for the Size of a Field

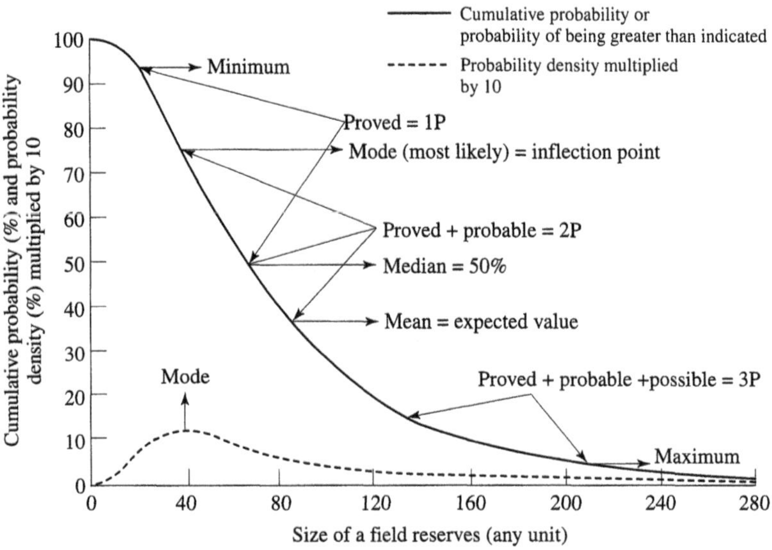

Source: Theoretical values

The term "proved reserves" is used in the US to satisfy the Securities and Exchange Commission (SEC) rules, often as a basis for financing, as they are said to have a "reasonable certainty" of occurrence. Various attempts to define "proved reserves" in probabilistic terms have been made. The professional bodies (SPE/WPC/AAPG and DTI) define proved reserves as having a 90 percent probability, whereas others define them as better than 50 percent. In practice, for the last 20 years there have been twice as many positive revisions of proved reserves in the US as negative ones. This implies a probability of around 65 percent (two-thirds), which corresponds in a lognormal distribution to the most likely (mode) case. It means in fact that the best estimate corresponds to the most likely case, not the maximum estimate. Failure to understand this may have contributed to a misplaced belief in the excessive future growth of reserves, which would have been greater if proved reserves did indeed have a 90 percent probability.

In short, proved reserves (or 1P) correspond with a probability varying from 50 percent to 95 percent; proved + probable (2P) with a probability varying from 40 percent (mean) to 65 percent (mode) and proved + probable + possible (3P) with a probability varying from 5 percent to 15 percent. For these reasons, proved reserves in the United States have been progressively revised upwards as some of the probable (and possible) reserves are upgraded to proved status when more information becomes available.

However, in the rest of the world, as in the North Sea, the reserves are given as proven + probable, as is done in the main industry database. In practice, technical estimates commonly differ from the reported values. Actors in the oil industry see advantage in under-reporting that allows a desirable image of growth, which is more important to them than the absolute values themselves. Such an inventory of under-reported reserves has often supported financial performance through years of lean discovery.

It is understandable, therefore, that most companies and countries prefer ambiguous definitions, or no definitions at all, in order to be able to report what suits their purpose best. Despite this laxity, many reports give a misleading impression of accuracy by reporting estimates to several decimal points. As the great mathematician Gauss commented long ago, a lack of mathematical thinking is obvious in the excess of accuracy in numerical calculation. This tendency is made worse by the advent of computers, which deliver calculations to a large number of digits instantaneously.

Since most oil data are accurate to within no more than 10 percent, only two significant digits should be given. An even worse practice is to convert rounded figures into detailed values in another unit, as for gas 1 Tcm $(T.m^3) = 35.3$ Tcf (instead of rounded 30 or 40 Tcf, using only one single digit as the original estimate).

Another bad practice is to write Gm^3 (cubic gigameter) for billion cubic meter, which is in fact a cubic kilometer, as $Gm^3 = Gm*Gm*Gm = 10^{27} m^3$ (around 10 times the earth's volume). If G is retained, for example to compare with Gt, it should be separated by a dot: $Gcm = G.m^3$.

Aggregation

Adding the estimates of a large number of fields is not straightforward, because the probability ranking of individual fields differs. Thus, the sum of the proved reserves of individual fields will be less than the proved reserves of a basin or country, as it is unlikely that every field estimate will be the minimum value. Only mean values may be added up. Yet the main public databases, as published by *World Oil*, the *Oil and Gas Journal* and BP, all mistakenly add up what are described as proved reserves.

This may be explained further by considering three fields with the following range of mini (proved) and maxi, as it is difficult to estimate the mean at first guess (see Table 1.1).

Table 1.1
Aggregation of Minis and Maxis

Field	Mini	Maxi
A	700	2000
B	32	100
C	1.1	2
Total:		
Usual practice	733.1	2102
Correct practice	750	2000

Source: Theoretical values

Reserves indicate potential future production. The uncertainty involved in assessing reserves in a basin is great, because in addition to geological and geophysical uncertainties, economic, technical and, not least, political uncertainties have to be taken into account. However, these genuine difficulties are small in comparison with those derived from lax and spurious reporting.

For most purposes this laxity does not particularly matter, but if one wants to use the record of the past to help forecast the future, one certainly needs to work with valid data.

Uncertain Production Data

Oil supply differs from demand by the change in stocks, as shown by the following data in Table 1.2.

Table 1.2
World Oil: USDOE/EIA Short Term Energy Outlook (August 2000)

	1999	2000
World demand Mb/d	74.8	75.8
World supply Mb/d	73.9	76.6

Source: USDOE/EIA (STEO, August 2000)

Table 1.3
US Oil Supply (1999)

	Mb/d	Percentage of Total Supply	Percentage of Domestic Production
Crude & condensate	5.88	30	76
NGL	1.85	9	24
Total liquids production	7.73	40	100
Imports	9.91	51	128
Processing gain	0.89	5	12
Withdrawal	0.3	2	4
Alcohol	0.38	2	5
Others	0.31	2	4
Total supply	19.52	100	253

Source: EIA

In the US, the importance of the NGL (24 percent) and processing gain (12 percent) versus domestic production is notable, as can be judged from the figures in Table 1.3.

The inventory of supply and demand is neither easy to determine nor reliable. The International Energy Agency (IEA) in Paris is generally regarded as the best source, but many errors have occurred. The exceptionally low price of $10/b in 1998 was due to the mistaken decision to increase OPEC quotas in the face of the Asian recession and a serious over-estimation of supply (300 to 600 Mb) by the IEA. It led many analysts, with a blind faith in market forces and technology, to conclude that the world had entered a long new chapter of low oil prices.

Oil supply comes from many different sources, each of which can contribute in a different way, but they are not distinguished properly in the statistics. Many reports do not explain in detail what they include, and values vary even in countries where official agencies such as the Norwegian Petroleum Directorate (NPD) in Norway or the Department of Trade and Industry (DTI) in the UK provide the data. For Norway (see Figure 1.2), the NPD gives 2.9 Mb/d for oil in 1999 and 3.1 Mb/d for oil and condensate, while BP gives 3.2 Mb/d.

For Canada (see Figure 1.3), the discrepancy is larger because of unconventional oils from tar sands. In 1999 oil production was 1.9 Mb/d according to *O&GJ* and the USDOE, 2.1 Mb/d according to the NEB (National Energy Board) for oil and equivalents, 2.2 according to *Petroleum Economist*, 2.6 according to BP and *World Oil*, and 2.9 according to NEB for all liquids. The range is as much as 1 to 1.5.

The world's present oil production is variously reported as either 66 Mb/d or 76 Mb/d, depending on whether only crude oil or all liquids are covered. The US Department of Energy (USDOE) database indicates the monthly production of crude oil by the main producers (see Figure 1.4). Crude oil comes from oilfields, but gas liquids come from both oilfields and gasfields. To overcome this difficulty it is better to study crude oil production separately where possible.

The non-OPEC countries have tried unsuccessfully since 1985 to raise crude oil production above 40 Mb/d. Claims are now being made that the

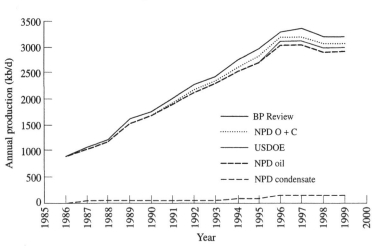

Figure 1.2
Norway: Oil Production Data

Sources: BP Amoco, NPD and USDOE

Figure 1.3
Canada: Oil Production Data

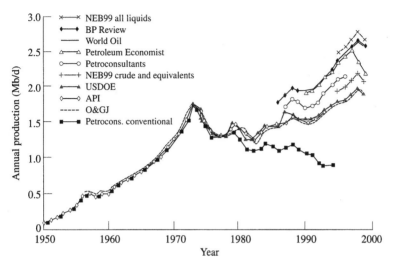

Sources: NEB 99, BP Amoco, *World Oil*, *Petroleum Economist*, Petroconsultants, USDOE, API, *O&GJ*

Figure 1.4
World Oil Production: USDOE Monthly Data

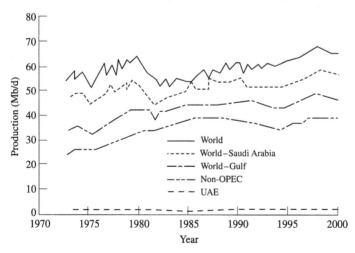

Source: USDOE

[25]

recent oil price rise and new discoveries in deep-water regions will make it possible to reach this level. However, such a production level is unlikely to be attained because of the high decline rate of old North Sea fields. In fact, the world outside the Gulf is producing at full capacity and still has great difficulty in exceeding 50 Mb/d. So far it has been the Gulf swing producers, and Saudi Arabia in particular, that have provided the necessary supply to meet demand. It is hard to know if they will be able to do so in future.

If annual oil production is uncertain, cumulative production is even more so, as the old records are poor, especially with regard to cumulative gas. The reason is gas flaring, which releases large quantities of gas in Nigeria and even in the North Sea. In 1999 2.2 G.m^3 of gas in the UK and 0.7 in Norway was flared. Lost oil is generally omitted from the production data. During the Gulf War in 1991 more than 1 Gb of oil was burnt in Kuwait, but it is not counted in most databases, although the reserves were reduced by a similar quantity.

Figure 1.5 shows the United Arab Emirates' cumulative oil discovery (2P) and cumulative oil production. In countries producing at full capacity, these two curves are usually parallel, but that is not the case for the UAE.

In this database the UAE cumulative oil production at the end of 1999 stands at about 20 Gb and remaining reserves at about 60 Gb. The UAE plot

Figure 1.5
UAE Cumulative Oil Production and Cumulative Oil Discovery

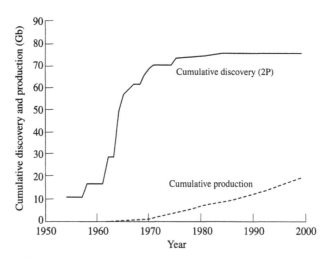

Source: Petroconsultants

[26]

Figure 1.6
UAE Annual Oil Discovery and Annual Oil Production

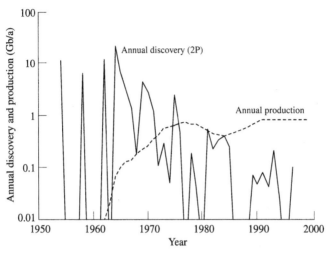

Source: Petroconsultants

for annual discovery and annual production shows that these two curves do not correlate, because production has been deliberately restricted (see Figure 1.6).

Discrepancies in Reported Reserves

Where reserves are kept confidential, it is not surprising to find different data from different sources, but it is surprising to find discrepancies in the official statistics of countries such as the UK and Norway (see Figure 1.7). This is due to the practice followed by US journals (such as *Oil and Gas Journal* and *World Oil*) of reporting reserves in accordance with reporting methods followed in their home country. As already discussed, the US reports on proved reserves fulfill the requirements of the SEC. In contrast, proved and probable reserves are normally reported elsewhere, even if, to still further confuse the matter, they are formally described as proved. The differences for the UK and Norway are very large between *O&GJ* and *WO*, both referring to 1P, and the industry database that refers to 2P. The latter database backdates revised estimates to the discovery of the field containing them, whereas in the US revised estimates of old fields are reported at the year of revision.

[27]

Figure 1.7
UK and Norway: Reported Reserves from Different Sources

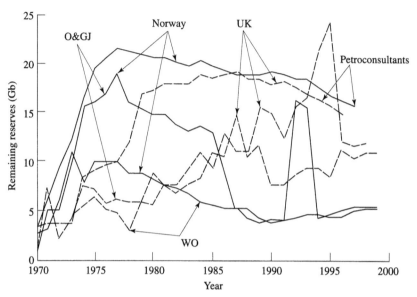

Source: Petroconsultants, *O&GJ* and *WO*

Since the UK and Norway presently produce about the same amount of oil (3 Mb/d), it seems obvious that the reserves should be about the same. According to *O&GJ* and *WO*, Norway's reserves are double those of the UK, but the reserves reported by the DTI and NPD are different (see Table 1.4).

Table 1.4
UK and Norway: Reported Reserves from Different Sources

	UK	*Norway*
Production 1999 (USDOE Mb/d)	2.7	3
Cumulative production end 1999 Gb	18	13
Remaining reserves end 1999 Gb		
O&GJ and *WO*	5	11
Industry database	9	12
NPD: crude oil		13
NPD: liquids		16
DTI: proven + probable	8.5	
DTI: proven + probable + possible	15	

Sources: USDOE, *O&GJ*, *WO*, NPD and DTI

Figure 1.8
UK and Norway Oil Production and Shifted Discovery

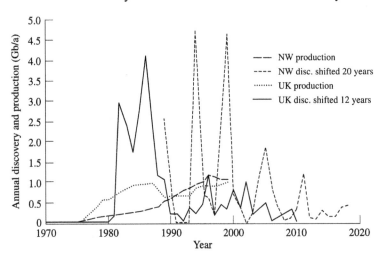

Source: Petroconsultants

UK production peaked in 1987, but there are two cycles of discovery, as shown in Figure 1.8. A twelve-year shift has provided a good peak correlation. It means that production in the UK will decline soon. Indeed, it may now be at its peak. For Norway, a twenty-year shift gives a good correlation and shows flattening in 1997 before a decline, unless a new discovery cycle occurs.

UAE Reserves

The remaining reserves of the UAE are reported at 98.1 Gb on the web, but vary from 60 to 100 in other sources (see Table 1.5).

Table 1.5
UAE Reserves

Remaining Reserves	Gb
DOE/EIA 1999	97.8
BP 1999	97.8
O&GJ 1999	99.6
WO 1998	63.5
USGS 1995	72
Industry 2P 1995	60

Sources: USDOE/EIA, BP Amoco, O&GJ, WO and USGS

The rise of 1986 reported by *O&GJ* and *WO* does not appear on the backdated curve on Figure 1.9.

Figure 1.9
UAE Reported Reserves from Different Sources

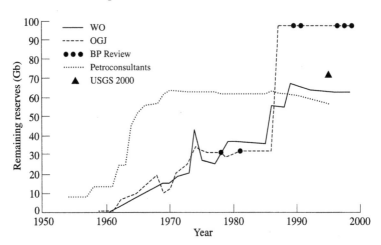

Sources: *WO*, *O&GJ*, BP Amoco, Petroconsultants and USGS 2000

Figure 1.10
Estimate of Abu Al Bukhoosh Reserves from Production Decline

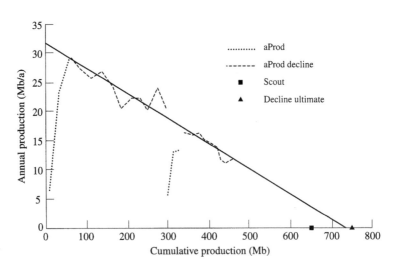

Source: Petroconsultants

The reported reserves per field obtained by industry intelligence reports are questionable, but the reports can be checked against production data from the field, when it is in decline and producing at full capacity. This is unfortunately not the case for most UAE fields. For example, industry sources report that the Abu Al Bukhoosh oilfield has 650 Mb of initial reserves, but decline analysis as shown in Figure 1.10 indicates a higher value of about 750 Mb. However, the production of 1987 to 1989 has to be discarded in order to determine a proportionate value.

Data on World Reserves

For the world as a whole, the remaining reserves (1P oil + some NGL) described by *O&GJ* and *WO* display an upward trend over the last 50 years, whereas the backdated 2P liquids from the industry database show a declining curve since 1980 (see Figure 1.11). Currently reported proved reserves mainly reflect poor reporting rather than the real status of discoveries. The reported proved reserves may please bankers and shareholders, but the real decisions of the oil industry are made according to different data, namely the mean values (approximated by the 2P = proven + probable).

Figure 1.11
Remaining Reserves of World Oil from Different Sources

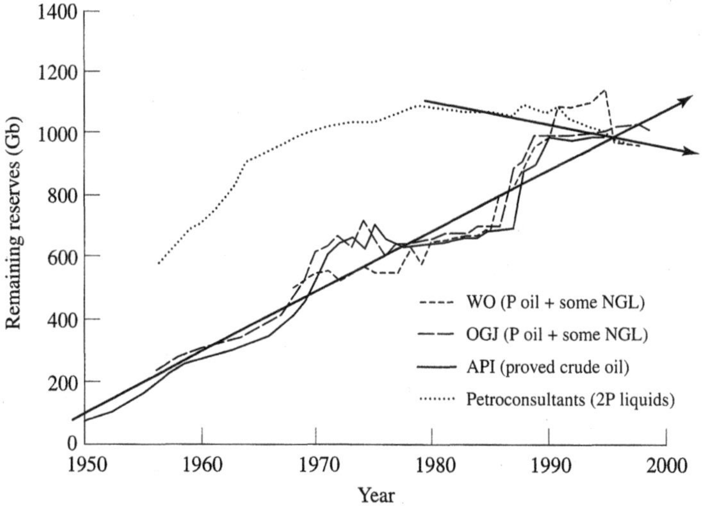

Sources: *WO, O&GJ*, API and Petroconsultants

[31]

In conclusion, it may be said that the quality of production data are poor and that of reserve data even worse. US Secretary Richardson, who tried to persuade OPEC to increase production in August 2000, requested better data. It is about time.

Methods: Means to Assess Potential

In a mature petroleum basin (and most of them fit this description) the potential for a new discovery may be assessed by several methods. The best method is to have complete data on past discoveries and good seismic coverage of the whole area to estimate the potential of every prospect and lead. However, it is rare to have access to such comprehensive data.

Estimates can also be made from comparing the volume of sediments or the area covered by a basin and its measurements with a similar known basin. This was the approach followed by Zapp of the US Geological Survey in 1961, when he estimated the total of US oil at 590 Gb, much more than Hubbert's 200 Gb. Zapp postulated (Hall 1992) that drilling 5 billion feet was required to properly explore the US, assuming one well to basement or 20,000 ft per 2 square miles (four times more than had been drilled until then). He also assumed that the recovery per foot would be the same as in 1961, which was 118b/ft, giving a total of $118 \times 5 = 590$ Gb. Hubbert objected at the time that the average recovery per foot was decreasing with time (it has been about 20b/ft for the last 30 years), pointing out that mineral (and oil) exploration is subject to the well-known law of diminishing returns, as the largest discoveries are made first, being too large to miss. Assessments based on sediment volumes have now been abandoned.

One of the best approaches is the creaming curve, initiated by Shell, which involves comparing the cumulative discovery (100 percent) with the cumulative number of wildcats (see Figure 1.12). Shell's record shows a beautiful smooth decreasing curve (close to a hyperbola) but consists of two cycles, the first ending in 1955. So far the company has found 60 Gb with about 4000 wildcats, but extrapolation shows that another 4000 wildcats would bring in only 20 Gb more.

Another approach is to study the field size distribution of natural domains with a common rock source, for example a petroleum system, using known discoveries from which to extrapolate the full system. The

Figure 1.12
Shell Oil Discoveries: Creaming Curve

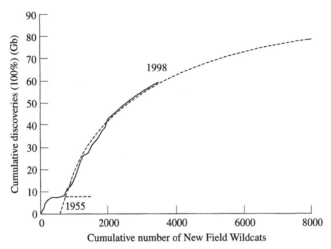

Source: Petroconsultants

distribution can be regular or fractal. A perfect regularity corresponds with a linear fractal (used formerly by the USGS and now by Robertson Research International), but in nature regularity is imperfect and the fractal is curved, close to parabolic. In this graph fields are ranked by size in decreasing order and then drawn according to prominence in a log format. The largest fields, which are indicated first, determine the parabolic parameters, which are extrapolated to draw the ultimate curve. The difference between the known discoveries and the total discoveries indicates the distribution of the undiscovered fields in the ground, which in turn indicates the potential reserves when time, economics and activity levels are taken into account. The fractal distribution may then be combined with the creaming curve to reach a sound conclusion.

UAE Potential

The UAE covers part of a petroleum system (PS) called the Rub Al Khali PS in the USGS 2000 study. This system belongs to a larger petroleum system called the Arabo-Iranian Mega-Petroleum System. As the UAE covers the bulk of the petroleum system, the fractal display is smooth and can be easily extrapolated (see Figure 1.13).

[33]

Figure 1.13
UAE Oilfields: Fractal Display in a Log-Log Format

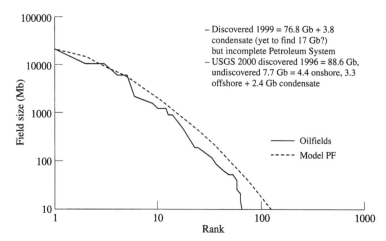

Source: Petroconsultants

Figure 1.14
Rub Al Khali Petroleum System: Fractal Display for the UAE

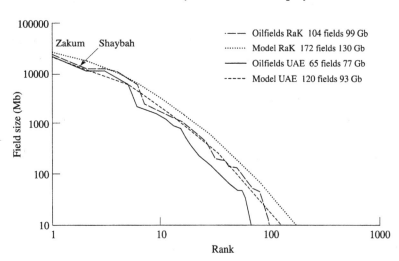

Source: Petroconsultants

The cumulative oil discovered, based on 1999 data, amounts to 77 Gb discovered in 65 fields of over 10 Mb each, and the total (140 fields over 10 Mb) is about 93 Gb. The fractal display for the UAE is not very different from that of the full petroleum system (104 fields over 10 Mb and 99 Gb) as shown in Figure 1.14.

Figure 1.15
UAE Oil Creaming Curve (1999 Data)

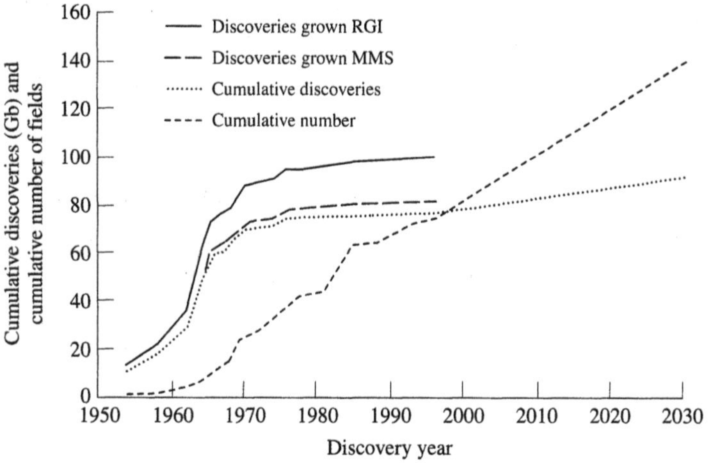

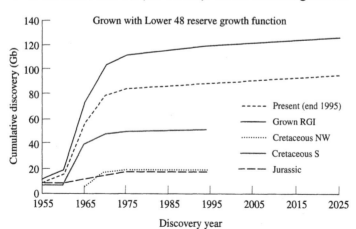

Source: Petroconsultants

Figure 1.16
UAE Undiscovered Oil (1996 Data): USGS Creaming Curves

Source: Petroconsultants

The UAE creaming curve is also regular, pointing to a total of 93 Gb in 2030. On this graph (see Figure 1.15) the size of the reported discoveries has been increased with the multiplier used by the USGS 2000 assessment for reserve growth in the world. The USGS admitted that it is impossible to obtain a reliable reserve growth function for the world as a whole and there-fore, in the absence of anything better, used the US Lower 48 mean reserve growth function. This function, called RG1, multiplies the reserves as follows: by 6 on discovery; by 1.6 ten years after discovery; by 1.4 twenty years later; by 1.3 thirty years later; by 1.25 forty years later and by 1 ninety years later. This curve is based on very old fields, mainly onshore, whose reserves were appraised with obsolete methods. The MMS estimates reserve growth for the Gulf of Mexico on the basis of more recent fields, which were better evaluated to begin with. It uses a smaller multiplier, being only 4.4 at the discovery year and 1 fifty years later. The cumulative discoveries (1999 data) are increased by 5.4 Gb on the MMS model and by 22 Gb on the USGS RG1 model.

Figure 1.16 is a graph of the USGS 2000 assessment values (1996 data) of the assessment units 20190101 (Cretaceous NW), 20190102 (Cretaceous S) and 20190202 (Jurassic). The total quantity discovered until the end of 1995

Table 1.6
Discovered and Undiscovered UAE Oil Reserves: Distribution by Size

Size Mb	Discovered		Parabolic Fractal	
	Number	CP Gb	Number	CP Gb
1000	12	68	15	80
100	35	75.3	48	90.9
10	65	76.8	121	93.3

Size Mb	Discovered		Yet to find	
	Number	Gb	Number	Gb
>1000	12	68	3	12.9
100–1000	23	7.3	10	3.6
10–100	30	1.5	43	0.9
>10 Mb	65	76.8	56	17.4

USGS 2000 undiscovered		7.7
USGS reserve growth RG1		30

Sources: Petroconsultants and USGS

was 89 Gb (CP = 15.7 Gb and RR = 73 Gb), which increased to 120 Gb. The USGS total for the UAE is more than 125 Gb (adding 30 Gb for reserve growth and 7 Gb for undiscovered oil).

Table 1.6 compares our estimate of undiscovered oil (17 Gb) with the USGS 2000 estimate, which is only 7.7 Gb, but with a field growth of 30 Gb and its Lower 48 mean reserve growth function applied to their 1996 database.

Robertson Research International (RRI) had undertaken a similar large study on the world's petroleum systems and presented it at the same time as USGS in Calgary at the 16th World Petroleum Congress in a paper entitled "World Conventional Hydrocarbon Resources: How Much Remains to be Discovered and Where is it?" by Richard M. Fowler. The world's undiscovered liquids, according to Robertson, amount to about 440 Gb, compared to the estimate by USGS 2000 of 939 Gb, but Robertson did not mention any reserve growth, while USGS 2000 adds 730 Gb, adding 1670 Gb for the next thirty years. This is 50 Gb/a, which is more than three times the annual discovery rate over the last ten years, an implausible figure.

Figure 1.17
Robertson Kimmeridge-Tithonian Carbonate Play
within the Rub Al Khali Petroleum System

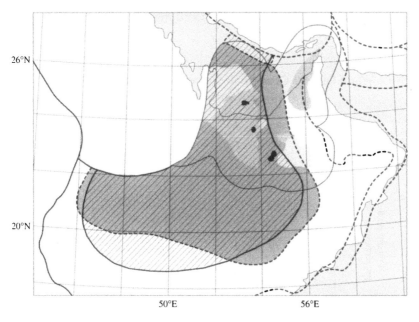

Source: Robertson Research International

Robertson uses a fractal display to assess the total but mistakenly cuts off the largest fields and draws a linear fractal. For example, Robertson estimates the undiscovered oil of the Kimmeridge-Tithonian Play (see Figure 1.17) at 6 Gboe (P50) with a range 3.4 to 9.5 Gboe, giving a total for the Rub al Khali of 89 Gboe (48 percent oil = 43 Gb for liquids). This is very optimistic compared with even the USGS estimate. Our parabolic fractal on Figure 1.14 gives 30 Gb of undiscovered oil and notes that gas liquids are important.

World Reserves

The different databases of conventional reserves (with different reserve definitions) have been combined to obtain a relatively homogeneous field database of near mean values, which implies no reserve growth. Figure 1.18 displays the conventional creaming curve extrapolated from these data, with a total of 2000 Gb. It appears that this total can reasonably be reached, with a 10 percent margin of error, in the next 40 years.

Our total estimate involved more than five years of study by four retired exploration geologists: Alain Perrodon, who was the first to introduce the term "Petroleum System," Gerard Demaison, who quantified the generation of a petroleum system, Colin Campbell and myself.

Figure 1.18
World Oil plus Condensate Conventional Creaming Curve
(with an ultimate of 2000 Gb)

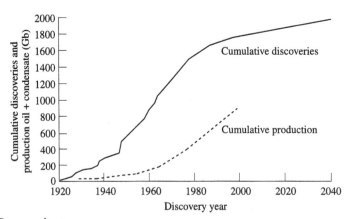

Source: Petroconsultants

[38]

Our total estimate is given in Table 1.7:

Table 1.7
World's Ultimates: Oil and Gas

	Mini	Mean	Maxi
Conventional oil	1700	1800	2200
Conventional gas liquids	200	250	400
Non-conventional liquids	300	700	1500
Total liquids (Gb)	**2300**	**2750**	**4000**
Conventional gas	8500	10000	13000
Non-conventional gas	1000	2500	8000
Total gas (Tcf)	**10000**	**12500**	**20000**

Source: Perrodon et al. 1998

The history of total estimates for the last 60 years is indicated in Figure 1.19. The average since 1960 is around 2 Tb.

Figure 1.19
Past Estimates of Conventional Oil and Gas Ultimates

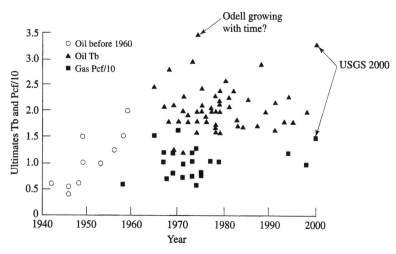

Source: Petroconsultants

[39]

Forecasting

Total recovery is a somewhat abstract concept indicating cumulative production when production has ended, and is of little practical interest unless it is properly related to a time frame. The best way of doing this is to use simple models and the best model is a symmetrical curve.

Forecast of Production up to 2050

The symmetrical curve in Figure 1.20 is based on the "Central Limit Theorem," which states that the sum of a large number of independent asymmetrical curves is a symmetrical (normal or Gaussian) curve. The problem is to obtain a large number of curves, which also have to be independent. The best examples are the oil production of the US (Lower 48) and the FSU, being the sum of a large number of fields, oil-producing companies and basins, which give symmetrical curves except when economic or political conditions, such as deliberate cuts in oil production or oil-price shocks,

Figure 1.20
US and FSU Oil Production and Forecasts with Hubbert Models

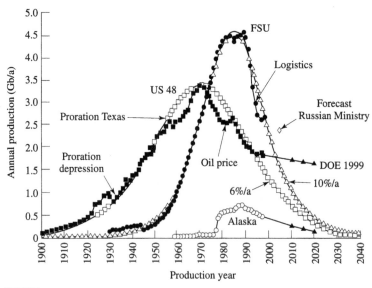

Source: USDOE

[40]

Figure 1.21
US (48 States): Annual Oil Production and Shifted Discovery

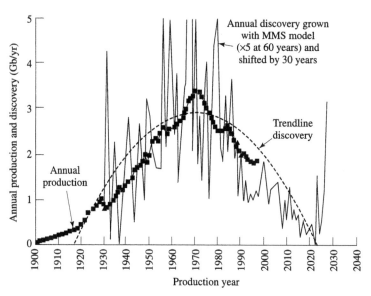

Source: Attanasi and Root, March 1994, grown with MMS model

disturbed the natural pattern. The USA as a whole is affected by a later second cycle that opened with Alaska.

In fact, after a certain lag in time, the annual production curve, when based on mean values, mirrors annual discovery. Figure 1.21 shows the considerable correlation between annual discovery (proved values, as corrected with the MMS curve to obtain mean values), shown for a period of 30 years, and annual production.

Oil Production: France, UK and Netherlands

When the number of fields is limited as in France, where there were two discovery cycles, the production curve displays two symmetrical curves (see Figure 1.22). The United Kingdom and the Netherlands also display two beautiful symmetrical production cycles.

[41]

Figure 1.22
France: Oil Production and Shifted Discovery with Two Cycles

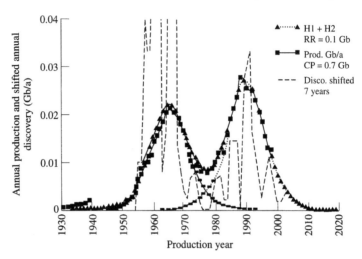

Source: Petroconsultants

UAE Oil Production

The UAE oil production curve peaked in 1977, but that was due to con-strained production and not to a previous peak in discovery. If the oil total is estimated at 83 Gb, with a cumulative production of 20 GB at the end of 1999, Hubbert modeling, shown in Figure 1.23, gives two scenarios of pro-duction without constraint for the next 80 years, with a total production of around 60 Gb for 2000–2080, but different peaks. These are just two possibilities.

Non-swing Producers

The swing producers are the five countries of the Gulf that have a swing role in making up the difference between world demand and what the non-swing countries can produce at full capacity. The correlation between shifted annual discovery (by 30 years) and annual production is striking; there is no need to draw a Hubbert curve to forecast the next 30 years' production, as the dis-covery curve tells it all. This graph (Figure 1.24) was shown at the last World Energy Council 2000 report: "Energy for Tomorrow's World – Acting Now."

Figure 1.23
UAE Oil Production and Hubbert Modeling

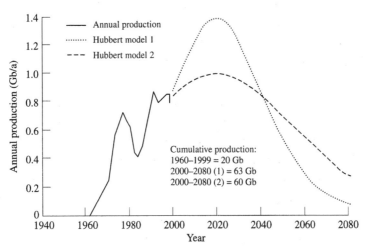

Source: USDOE

Figure 1.24
Non-swing Production and Shifted Oil Discovery

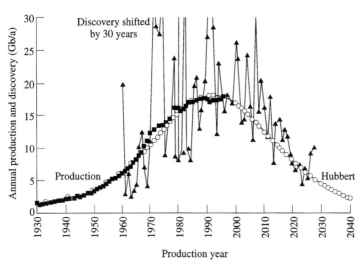

Source: Petroconsultants

World Forecasts

Unfortunately, the forecasts by official agencies, such as the USDOE, show a rising trend of oil production, without revealing when the peak and decline of production will be (Figure 1.25). They tacitly assume that oil resources are infinite. The forecast for future non-OPEC production is certainly optimistic when related to past experience.

The USDOE 1999 forecast of future oil prices (Figure 1.26) is also very optimistic when it suggests that the price of oil will be essentially flat in the future, with a price of $22/b (1998) when world oil supply is at 113 Mb/d in 2020. The DOE was in fact required to correct the 1999 forecast, but it assumes that the 2000 price of $25/b will be an anomalous spike of very short duration. In fact, the present oil price is in marked contrast with this DOE forecast.

In Figure 1.27 we have drawn the minimum curve extending up to the end of production to fit with the USDOE/EIA, International Energy Outlook (IEO) and IEA liquids forecasts for 2020, in order to calculate the total implied by these forecasts. This minimum curve needs 2600 Gb for the IEA forecast, which is close to our total for unconventional oil, and 4500 Gb for the DOE/EIA forecast, which is much higher than the USGS 2000 total of

Figure 1.25
World Oil and Liquids Production and Forecast by USDOE

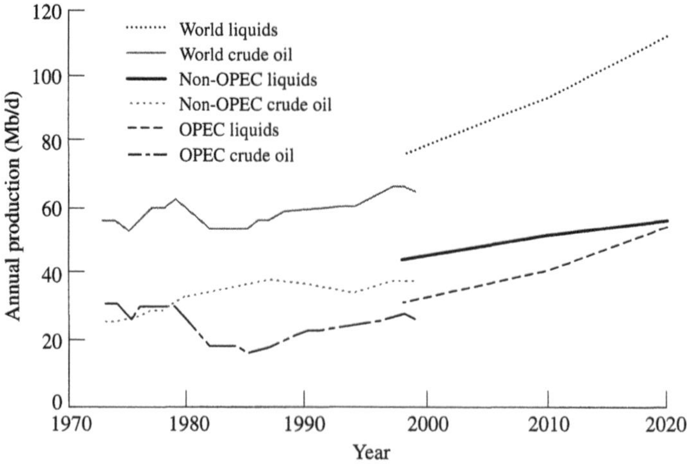

Source: USDOE

Figure 1.26
USDOE Oil Price Forecast

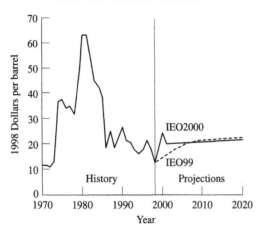

Source: USDOE

3003 Gb. This forecast of 3003 Gb, which, incidentally, should certainly be rounded to 3000 Gb, was accepted by John Wood of the USDOE/EIA in his 2000 paper on long-term supply. He proposes twelve scenarios, from which we have selected the scenario with 2 percent per annum growth until 2016 followed by a 2 percent per annum decline, and the scenario of 2 percent growth per annum until 2037 followed by a decline of R/P = 10 (meaning a decline of 10 percent per annum). This second scenario agrees with the forecast of EIA/IEO 2000, but the strong decline is unrealistic, both in terms of natural decline and the interests of future generations.

The problem is that much of the database and study of the USGS is based only on conventional oil, while significant amounts of non-conventional oil and gas liquids are already being produced. Only scenarios covering all liquids give the global picture of supply and demand. Conventional forecasts can only address (as did our article in *Scientific American* in March 1998) the end of cheap oil, which is by no means the same as all oil.

Figure 1.28 (Laherrère 1999d, *CSEG*) gives our scenario three symmetrical curves for conventional oil, unconventional oil and conventional + unconventional gas liquids, with a total of 2700 Gb. The peak is reached around 2010 as the production of unconventional oil (mainly Orinoco and Athabasca) can increase only slowly, save perhaps in response to a radical increase in oil price.

[45]

Figure 1.27
World Oil and Liquids Production: USDOE/EIA and IEA Forecasts
(fitted to Hubbert Models and Ultimates)

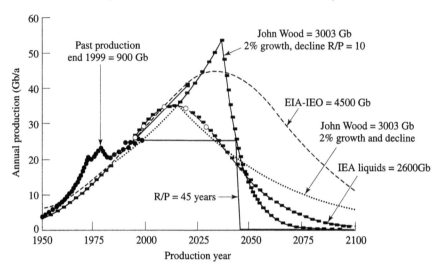

Sources: USDOE/EIA 2000 and IEA 1998c

Figure 1.28
World Liquids Forecast

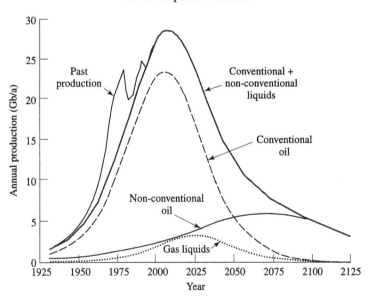

Source: Perrodon et al. 1998

[46]

Figure 1.29
World Population Scenario and Oil and Gas Production per Capita

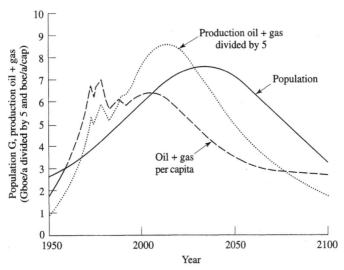

Source: Perrodon et al. 1998

Per Capita Oil and Gas Consumption

In short, we expect that global oil and gas production will peak within the next 20 years. In this connection, we may also examine the issue of per capita oil and gas consumption. The UN forecasts of world population have decreased since 1990, when the global population increase peaked at 90 million per year, before falling to its present level of 78 million per year. The world's population can be modeled with three Hubbert cycles for industrialized countries, developing countries and countries with limited education, such as Afghanistan.[1]

Figure 1.29 displays a scenario with oil and gas consumption peaking around 2010–2020 and population peaking around 2030–2040. In fact, oil and gas consumption per capita peaked in 1979 at around 7 boe and it will peak again soon at around 6.5 boe. It means that there will be an imminent and irreversible decline of oil and gas consumption per capita.

Alternatives to Oil: Liquids

Although the world is approaching the close of the cheap oil phase, this does not spell the end of oil as a source of energy, especially for transport. There

Figure 1.30
Primary Energy Consumption per Capita: IEA

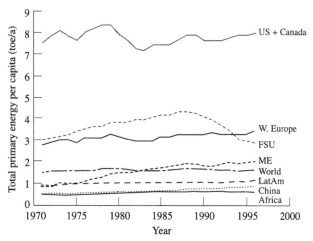

Source: IEA

are several ways in which to compensate for the decline in oil production. Saving energy, condensate, NGL, synthetic oil and gas-to-liquids are some of these ways.

- Saving energy would affect the Sport and Utility Vehicle (SUV), so popular in America, which could be replaced by more economical cars. Americans use twice as much energy (8 toe/a = 60 boe/a) as Europeans (3.5 toe/a) and could therefore easily achieve energy savings (see Figure 1.30), given that the world as a whole uses only 1.7 toe/a (12 boe/a /cap) (with Africa using just 0.6 toe/a).
- Many gasfields recover the condensate, while the gas is re-injected into the reservoir both to stimulate oil production and for eventual recovery. In the case of Prudhoe Bay, for example, condensate amounts to 700 Mb, being 5 percent of the field's total liquids. The world's largest gasfield is made up of the North Field of Qatar and the South Pars of Iran, together containing over 30 Gb of condensate.
- The US produces several times more natural gas liquids (NGL) than the rest of the world, but with more plants the rest of the world can recover more liquids, which are presently flared with the gas.
- The Fischer-Tropsch process can make hydrocarbon liquids or synthetic oil from coal, and was used by Germany during World War II and by

[48]

Sasol in South Africa during the period of economic sanctions, but it is not profitable in the present environment.

- GTL (gas-to-liquids): There is at present only one pilot plant making synthetic oil from natural gas, namely Shell's Buntulu plant, which produces 12,000 b/d from 120 Mcf/d (1 boe = 10 kcf), but it is an old plant that is uneconomic. Many projects are under way but profitability is good only where the gas is cheap or even at a negative cost in places such as Nigeria, where flaring gas is taxed.

- There are three separate gas markets: Europe, Asia and North America. The price of natural gas in North America is soaring due to the sharp depletion of the gasfields and a failure to replace them by enough new gas wells. Huge gas reserves are still undeveloped in other parts of the world but are too far from markets. GTL will be expensive and located only in cheap gas areas.

- Alcohol from biomass, such as ethanol from corn and sugar cane, is another possibility. It seems that the net energy of the process is nil or negative, so that such projects depend on subsidies.

- Oil shale is formed by immature source rocks and would be better classified as coals rather than hydrocarbons. They may be used as a direct source of heating, for example in the FSU (Estonia). To produce oil, they require pyrolysis at over 500°C. Oil shale plants were built in France (Autun) and in Scotland around 1850, and the first extraction of oil from oil shales is reported in China around 1700. There are huge deposits of oil shale, mainly in the US, but most of the projects undertaken both by mining and in-situ combustion over the past 80 years have been failures, despite subsidies. They are uneconomic and likely to have a negative net energy yield.[2] They also need much water and produce more fine grained residues than it is possible to handle. In 1980 the President of Exxon claimed that the US had resources equal to 1 Tb of oil in the form of oil shales, enough to supply the country with 15 Mb/d for 175 years. The USDOE had forecast that oil shale production would reach 2 Mb/d in 2000, but the investments failed. There is a new project in Australia, namely the Stuart pilot project, built between August 1997 and April 1999, but it has been delayed by odor problems. The cost of Stage 1 is $250 M for 4500 b/d, but requires an investment of $50,000/b/d when deep-water oil yields less than $10,000/b/d. So far oil shales are decidedly uneconomic.

Alternative Energy

Alternative energy resources include hydrogen and unconventional oil, for example coalbed methane, tight sands, gas shales, geopressured aquifers and hydrates.

- *Hydrogen*: This is often cited as the fuel of the future, but hydrogen is not in itself a fuel, because its production depends on fossil fuels (reformed gasoline, methanol, ethanol or natural gas) or electricity by electrolysis from nuclear plants, hydroelectric plants, windmills or solar panels.
- *Coalbed methane (CBM)*: production has increased in the US but the rate per well is low, meaning that it is mainly of interest as a local supply. However, China and Russia have considerable potential.
- *Tight sands and gas shales*: These are also characterized by large resources but small productivity.
- *Geopressured aquifers*: A huge amount of methane is dissolved in these aquifers, but pilot fields in the Gulf of Mexico were uneconomic because they faced many environmental problems.
- *Hydrates*: Oceanic gas hydrates are unlikely to be economic.[3] Rather like gold in seawater, these are dispersed with low concentrations (less than 5 percent in Blake Ridge leg 164). Collett (USGS) first announced hydrate resources as being twice as much as all conventional fossil fuels (700,000 Tcf), but has now revised his estimate downward, claiming that they equate to no more than conventional gas (5,000 Tcf). No reliable method of producing oceanic hydrates has yet been developed and the obstacles are great. Gas hydrates occur also in permafrost, but the reported production in Messoyakha gasfield is questionable.[4] Recently the Japanese National Oil Company (JNOC) drilled a well in the Nankai Trough through 950 m of water to investigate oceanic hydrates. The well reportedly found hydrates, but since then there has been no further news. Whatever was found is probably insignificant. Hydrates are well known as an operational hazard, clogging tubing and pipelines. However, they may at the same time offer a means of transporting conventional gas more cheaply in the form of hydrates rather than as LNG.[5]

Conclusions

The main challenge in assessing the world's future oil supply is overcoming the poor quality of the data on both production and reserves, which is subject to much "political" interference. The best hope of remedying this critical issue is for the national oil companies or agencies to form an apolitical organization to agree on definitions and provide valid data. In many countries the truth may prove to be more powerful than lies. Therefore, such a policy could have many political advantages. Such a proposal may sound utopian, but it is the only way to gain a better understanding of this critical subject, which is of vital interest to mankind.

There is a large range in the total estimates, but for the most part they relate only to conventional oil. Most of the world's oil endowment has already been found. The peak of discovery in the 1960s will undoubtedly be followed by a corresponding peak of production within the next 10 years.

After 2010 the Gulf and the unconventional oils of the Orinoco and Athabasca will be the main sources of supply, although gas liquids from remaining gas reserves, which are less depleted, will also make an important contribution. The oil price will be high in order to balance demand and supply to a more reasonable level of production than the one presently forecast by most official agencies. The transport industry will call for more natural gas. In contrast, the future of coal is more difficult to assess. Renewables will increase, but without subsidies the net energy of most processes will temper production. Nuclear energy supply will have to increase to satisfy the demand for electricity.

ALTERNATIVES TO OIL

2

Alternative Energy Sources:
An Assessment

David Hart

Few true alternative energy "sources" have escaped detailed analysis. There is, however, a growing realization that a combination of energy sources, conversion technologies, economic pressures and environmental pressures needs to be considered holistically to enable further understanding of the future prospects of energy.

This chapter considers a number of "conventional alternatives" as energy sources, for example, solar power, biomass and wind power. It does not analyze the potential for wave power or other more esoteric options, although these may become important at some point in the future. It also does not consider nuclear power or hydroelectric generation.

The reasons for the latter omissions are straightforward. Nuclear power is an accepted mainstream technology at present in many countries, though no new nuclear plants have been ordered in the United States, for example, for 2 decades. In a majority of countries it is either uneconomic or heavily subsidized, and market liberalization is rendering even high-risk investors wary of investment in new nuclear plants that have very high up-front capital requirements and long construction times. The risk associated with disposal costs is also too high. While there may be future developments in technology that allow nuclear power to compete on a smaller scale, this possibility is outside the scope of this chapter.

Both large and small-scale hydroelectric generation is technically well understood. While large-scale schemes such as the Narmada Project in India and the Three Gorges Project in China are deeply controversial, hydropower cannot be considered an "alternative" form of generation.

Instead, the chapter seeks to understand some of the drivers of change within energy markets, from an economic, environmental, political and technical perspective, and to use these to consider both energy demand projections and possible alternative future energy carriers. In some cases, these drivers will directly influence demand for petroleum products; in others the effect may be more subtle or negligible. From a geographical perspective, it is clearly progress in the developing world, particularly Asia, which will have the major impact on future energy.

The chapter closes with an indication of the changes that may occur in energy use, and where and why they may happen. The suggested twin future energy carriers of electricity and hydrogen could link together fossil and non-fossil energy structures, but also provide the potential for a move from one to the other.

Drivers for Alternative Energy Sources

While the world's use of energy continues to grow, and will grow faster over the coming decades, a complex mixture of drivers is affecting the primary resources that will be used, regional specializations, and the preferred energy carriers of the future.

These drivers – economic, environmental, political and technical – will have a long-term impact on issues such as pollution, import dependence, energy equity and local development, and could potentially have a significant impact on world trade. Drivers can be considered to be both positive (innovation and forward-thinking development) and negative (regulations on emissions reductions, resource depletion). There are also broadly neutral drivers such as those associated with urbanization and mechanization. However, all of them may be converging to push future energy use in the same direction.

Influence of Drivers

Throughout history the most substantial drivers for change in energy use have been technological. The development of the steam engine significantly increased demand for coal, while the internal combustion engine fed on newly discovered oil. Gas turbines and uses for the natural gas found in oil-fields developed in tandem.[1] The discovery of electricity and its many uses also had a profound effect. However, intertwined with these technological developments have been a series of negative drivers, primarily associated with pollution and scarcity of resources. For instance, the old-fashioned smog in London resulted in stringent legislation on coal burning.[2] In terms of current developments, it is the combination of these drivers that is particularly important.

"Negative" Drivers

The "negative" drivers of energy use are comparatively clear and they stem from concerns over environmental issues, equity, resource availability and security. As such, they can be categorized and explored.

Emissions: Local Pollution

The first and strongest driver of changes in the use of energy, whether in respect of types of energy or control of emissions, is immediate personal risk. Airborne smoke pollution, sickness, crop damage and building damage are all very tangible and can provoke strong reactions in those who may not otherwise understand energy issues in any depth. These phenomena also have the power to influence politicians.

It is for these reasons that regulations on vehicle emissions, in particular in OECD countries, have become so strict, as shown in Figure 2.1. The result is that, although the number of vehicles in OECD countries is rising, along with the number of miles they travel, the amount of severe pollution that they cause is dropping slowly. It is estimated that air quality in urban areas will continue to improve to meet World Health Organization and local guidelines, as future and more stringent regulations are introduced.

The most publicized manifestation of this driver is in California, where the Air Resources Board (CARB) has introduced the toughest regulations in the world. The zero emissions vehicle (ZEV) mandate will come into effect in 2003, by which time all automotive manufacturers with a certain market share will have to make a number of ZEVs available for sale to the general public. In total, 10 percent of their total sales must be comprised of ZEVs,

Figure 2.1
Evolution of EU Emissions Standards for Light Duty Vehicles

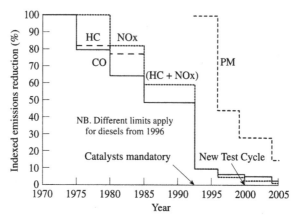

Source: EC Directives

but in the early stages of the mandate it is possible to fulfill the quota by sell-ing 2 percent "absolute" ZEVs, 2 percent hybrid-electric vehicles and 6 percent very low emission vehicles (VLEVs). The former must either be battery-electric vehicles or direct hydrogen fuel cell vehicles (FCVs); the latter may include hybrid vehicles, methanol- or gasoline-based FCVs, or meet some other specific requirements.

It is not clear how the manufacturers will meet these quotas, though General Motors (GM) announced in early 2001 that it was suing CARB over the continued requirement for ZEV adoption in 2003. Nor is it obvious how the general public will react once they are able to purchase the new vehi-cles, though the public response to the mandate itself has been overwhelmingly positive. However, there is a significant amount of devel-opment under way into FCVs, in particular, with the multi-company California Fuel Cell Partnership designed to give some exposure to the first sets of vehicles.[3]

Emissions: Regional Pollution

Regional pollution such as acidification, though marginally less obvious than urban pollution, is nevertheless a strong driver for regulation and for changes in energy use. The realization that acid rain, primarily from coal-fired power station emissions, was severely damaging woodlands and aquatic life in Germany, Scandinavia and North America led to the development of the Convention and Resolution on Long-Range Transboundary Air Pollution governing emissions of acidifying pollutants. There are now also more specific national and international regulations governing the emissions of the sulfurous and nitrogenous compounds that cause acid rain, and this has even led to the development of new currencies in that, for example, in some places it is possible to trade permits for sulfur. The use of coal and, in some cases, oil for power stations has been made noticeably more expensive by requirements for desulfurization at the flue, and this has been partly responsible for a trend away from higher carbon-content fuels toward lighter ones such as natural gas.

Emissions: Climate Change

While there is no irrefutable evidence yet of climate change caused by the accumulation of greenhouse gases (GHGs) in the atmosphere, the strong consensus among climate scientists is that it is occurring. The Intergovernmental Panel on Climate Change (IPCC) stated in its Second Assessment Report: "The balance of evidence suggests a discernible human influence on global climate."[4] The message was further reinforced by the Third Assessment Report, which suggested that "there is new and stronger evidence that most of the warming observed over the last fifty years is attributable to human activities."[5]

This is probably as strong a message as could be expected to come out of a conservative, politically sensitive, international organization, and suggests that there may well be significant climate change issues to confront in the future. In any case, the concerns that have been aired are being acted upon both in the international political arena and, more interestingly, in the international corporate sector. Policy makers have proposed, if not ratified, the Kyoto Protocol on national reductions of a group of GHGs, while major corporations such as BP and Shell have made internal commitments to reduce their own emissions of CO_2.

In addition, it now appears that both governments and corporations are prepared to initiate action along the lines of the "precautionary principle." Simply stated, this suggests that it is prudent to try to reduce GHG emissions before a causal link to climate change can be proven, as once it has been proven, it may be too late. Full ratification of the Kyoto Protocol is unlikely in the near future due to an adverse political climate in the United States, but it is possible that the EU, Japan and Russia could ratify and bring the process into force without US participation, and certainly there is significant activity in Europe regarding carbon-trading schemes and how they might be implemented. The process has been thrown into greater uncertainty by President Bush's announcement in early 2001 of his government's absolute rejection of the Kyoto Protocol.

Resource Availability: Scarcity Issues

Of crucial importance to the future balance of energy use are pressures from the clear and recognized dominance of fossil fuels within all sectors of the economy of most countries, other than in special situations such as in Norway, where most electricity supply is from hydroelectric schemes. The first pressure is simply defined as scarcity: while there is no immediate sign of overwhelming depletion of fossil resources, it is inevitable that a combination of the rising demand for all forms of energy and the ultimately finite nature of these resources will lead them to become scarce at some point in the future. This will clearly affect supply and price structures, and is likely to enhance the competitiveness of alternative sources and routes. This effect may even occur without any actual scarcity, for example if oil production is constrained to enable higher prices to be maintained or if policy measures are introduced to combat a perception of scarcity.

Resource Availability: Import Dependence

The second, and in many ways more important, pressure regarding resource availability is much more political than market-driven. Countries with limited fossil resources are heavily dependent on imports of petroleum products, coal, natural gas and other forms of energy, and are aware that their economic position is thus somewhat outside their own control. In addition, significant political capital can be made in some countries from a perception of "self-sufficiency."

Many countries thus state explicitly in their energy policies that they intend to reduce import dependence by fuel switching, demand-side management programs and other initiatives. The EU, worried that its current level of dependence at 50 percent may increase to 70 percent during the period to 2010, intends to increase the penetration of renewably generated energy from 6 percent to 12 percent during that time, while also increasing energy efficiency.[6]

"Positive" Drivers

Though they may ultimately have a similar effect on future energy trends to the "negative" drivers, the "positive" drivers of changes in energy use stem from different considerations.

Technology Development and Innovation

As discussed above, there has been a significant link in the past between technology development and energy use, and innovation still plays a very important role in the development of future energy scenarios. Significant development of renewable energy technologies such as wind power systems, solar photovoltaics and biomass systems has already taken place and is continuing, while the fuel cell is likely to play a major role in future energy systems, as will be discussed later.

Important though technology is, government policy in this area has a significant effect on both the development of the technology itself and its uptake into the market. Aggressive development and strong policy support in Denmark aided the development of a wind turbine industry, while Britain, with greater wind resources, was unable to achieve the same focus.

Perhaps more important for the future is a consideration of how early investment in research and development of technology can pay back in the future. The introduction of new technologies into carbon-reduction scenarios under uncertain conditions can be modeled using a Monte Carlo analysis, showing that early investment almost always results in a better payback and increased performance than late investment.[7]

Market Liberalization

Energy markets throughout the developed economies are being liberalized, with electricity and gas markets in various stages of re-regulation in many different areas. This trend is providing considerable impetus for alternative means of providing services to consumers, with a strong indication of moves toward the decentralization of electricity generation and supply. The immediate impact of this on oil markets is negligible, but natural gas suppliers are under significant pressure in markets where it is now possible to buy from a range of providers.

Nevertheless, the trends should not be ignored. It is possible to trace a number of drivers that may have significant impacts on future oil markets, brought about by decentralization. Consider the partially futuristic scenario that follows:

- Increasing congestion in electrical grid systems (where supply cannot match demand due to inadequate infrastructure) leads to widespread introduction of small-scale fuel cell systems at substations and in local distribution networks. This brings down the unit cost of fuel cell systems and brings wide public recognition.
- The fuel cell system is reversible (it can produce electricity from hydrogen or hydrogen from water using electrolysis) or can be coupled with an electrolyzer. This allows hydrogen to be produced when electricity is cheap and to be consumed to produce electricity once more when the price is higher. It also provides hydrogen that can be used in other ways.
- Availability of hydrogen, coupled with developments in fuel cell vehicles and the construction of a small refueling network for fleet vehicles, increases the attractiveness of direct hydrogen fuel cell vehicles, resulting in significant uptake.
- Hydrogen, produced from sources other than oil, begins to displace petroleum products from the transport sector.

The trends toward decentralization also affect the potential for future vehicle fueling. If, as has been suggested by some studies,[8] hydrogen can be produced cost-effectively at the fueling station to develop a refueling infrastructure for FCVs, then petroleum products will begin to be displaced as the first introductions are made. However, no significant impact on demand for gasoline and diesel within a country or region is likely for 10 to 15 years after the introduction, as the initial percentage of alternative fueled vehicles will be low.

Long-term Strategies

Some evidence exists that countries are considering longer-term strategic energy developments than those typically limited to a single election period. Germany is undertaking a "Transport Energy Strategy Study" (TES), with the specific objective of finding an alternative to petroleum-based fuels for the future.[9] The study has the full participation of Shell, DaimlerChrysler, BMW, Aral and the German government, amongst others, and more companies are planning to join. As yet no final decision has been taken, but indications are that the choice rests between natural gas, methanol and hydrogen. The European Union (EU) also has a strong interest to enlarge the study to encompass the whole of the EU.

While no similar development exists in the United States, there is a study under way in Japan that seeks to accomplish similar aims. Japan is in a unique position as it imports all of its transportation fuel, and the importance of supply diversity and consideration of alternatives is thus very high on the agenda. Clearly, should both the EU and Japan decide to introduce a particular alternative fuel, then this will trigger a demand for vehicles and fuel that will not use petroleum and may even have a significant influence on global automotive manufacturers.

The only comparable driver in the US comes about for totally different reasons. The California ZEV mandate discussed earlier is not derived from concerns over long-term dependence on oil, but from concerns over long-term polluting emissions. Its effect, however, should the only vehicles acceptable under the ruling be hydrogen FCVs, could be similar.

Moving to Sustainability

The concept of sustainability is not clearly defined, though E.F. Schumacher's concept of living off the interest from natural resources, not the capital, gives a good indication of the notion.[10] However, it is clear that some organizations are trying to understand the concept and are adopting measures to enhance their performance by following some sustainability principles. An example is BP, which appears to be trying to move beyond rhetoric and embrace both environmentally responsible performance and low-carbon futures. Whether this can be achieved, and which direction this will take, will depend both on its deliberations over the coming year or two and on movements in energy and related issues in the near future.

"Neutral" Drivers

Another set of drivers can be categorized as "neutral," stemming from changes in infrastructure, modes of working or social factors. These drivers will not be considered in detail but include the role of economic growth and structural change, and of energy prices. The former is a broad category, including changes in industrialization, mechanization and information technology (IT), with switches from industrial to service industries having potentially important consequences for energy demand, particularly for the type of energy demanded. As an economy develops its IT capacity and moves away from industrial dependence, demand for more organized forms of energy such as electricity is likely to increase, while other forms of energy may be less important. Urbanization will also have an impact on the type and quantity of energy required for an economy to function.

Energy prices, and the difference between these prices for different types of delivered energy within an economy, will have a significant impact on rates of growth and the types of energy that are favored.

The Effects of the Drivers

The drivers outlined above will each have specific effects on energy use within the geographical areas in which they apply. They may also, like the California ZEV mandate, spur innovation and development in multinational corporations, and thus have a sphere of influence that is well beyond their immediate geographic limits. In some cases the drivers will combine, and it may be that the effect is greater than might be expected from any one driver taken in isolation. One example is the development of vehicles that may address the ZEV mandate in California, which is exclusively devoted to criteria pollutants, while also meeting the voluntary agreements made between the automotive manufacturers and the EU regarding CO_2 reductions from vehicles. The latter agreement states that the manufacturers have to reduce the average emissions from their vehicles sold in 2008 by 25 percent compared to current levels: from 185 g/km to 140 g/km.[11] By 2010 they have to achieve 120 g/km. This can be done by the introduction of lighter, smaller models but the general trend is away from this, with increased on-board accessory requirements and heavier vehicles associated with greater safety. It may therefore be easier for the manufacturers to introduce a greater number of alternative fuel vehicles with very low CO_2 emissions to improve the fleet average. This may only serve to move the source of the emissions, as the upstream fuel production does not appear to be considered in the figures.

Whatever the effects of these specific pieces of regulation, the impact of the Kyoto process is already beginning to show. Shell and BP have internal carbon emissions trading schemes, and the World Bank has also set up some pilot work. The UK government has an Emissions Trading Group investigating the potential opportunities, barriers and pitfalls of national and international implementation of the concept. Large corporations are clearly beginning to move toward understanding lower-carbon futures and implementing measures to move further.

The effect of all this is not yet clear. Reducing effective carbon emissions to the atmosphere can mean very different things, from increasing carbon sinks such as forestry, to deep-ocean carbon sequestration of CO_2 from fossil fuels, to a fully renewable energy future. It is likely that all of these things will play a part. It probably also means a gradual movement away from petroleum-based fuels in the economies that rely most heavily on them – the OECD countries. What impact this has on global demand is not clear.

Future Scenarios for Energy Use

All detailed projections for energy use show large future increases, mainly driven by population growth and increasing standards of living in the developing world. The sources and supply of energy are subject to significant uncertainty but become generally more diverse over time. One message seems to be that growth in energy use in the developing world will have to be supplied largely by non-fossil sources. However, the details of these sources are unclear and their penetration will take a considerable period of time. Nuclear power is included in some scenarios.

The developing world will be the largest future energy market, growing rapidly from its current low level, particularly in Asia. Supplying that market while responding to the drivers discussed earlier in this chapter will define the influence of different technologies and primary energy sources in the future.

Projections for Energy Use

Many organizations have produced some projections of energy use, including the International Energy Agency (IEA), Energy Information Administration (EIA), Shell and the International Institute for Applied Systems Analysis (IIASA). The IIASA figures have the longest timescale, reaching to 2100, and also include scenarios driven by different assumptions – with wide variations in final demand and in supply mix.

Two IIASA scenarios are shown below – scenario B is a "business as usual" analysis, while scenario C1 is severely constrained by environmental considerations, with the result that the final energy use is far lower and the supply mix less fossil-dependent. Figure 2.2 shows scenario B by primary energy use, while Figure 2.4 gives the same scenario broken down by region. Figure 2.3 and Figure 2.5 show the same for scenario C1. The detailed assumptions underlying the scenarios are available in Nakicenovic et al.

Figure 2.2
Projection for World Primary Energy Use to 2100
under Business as Usual

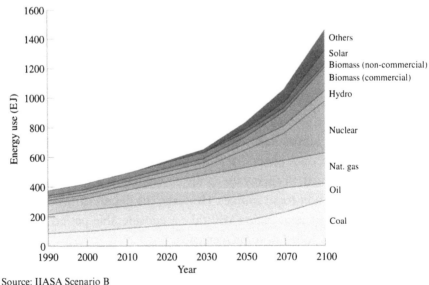

Source: IIASA Scenario B

(1998)[12] or at the IIASA website[13] and serve here to illustrate some of the issues under consideration.

Although a number of conclusions can be drawn, what is immediately clear is that for the "business as usual" scenario, oil use continues at approximately the same levels as today for at least 50 years. Even in the strictest environmental case there is no trend away from oil until after 2010, and until 2050 there is still a significant level of demand. Alternative sources of energy, unless nuclear energy is considered (and in this scenario there are some significant assumptions that waste issues have been resolved and smaller-scale nuclear technologies made available), provide less than 40 percent of the final total. This is more than the 400 EJ of current demand but still a very small proportion.

More interesting for the future of alternative energy sources is the environmentally aggressive scenario, C1. In this example there is a 40 percent contribution by the end of the century from solar energy alone, but also a widespread adoption of biomass, each starting in about 2030. While solar energy covers many options, this scenario assumes a considerable uptake of photovoltaic systems, and thus there are important questions that arise

Figure 2.3
Projection for World Primary Energy Use to 2100
under Environmental Constraints

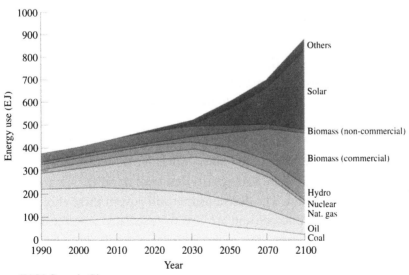

Source: IIASA Scenario C1

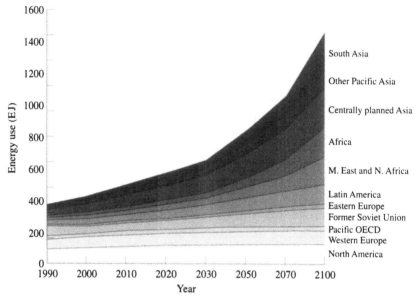

Figure 2.4
Projection for World Primary Energy Use to 2100
under Business as Usual by Region

Source: IIASA Scenario B

regarding the use of electricity. Can it be used directly? Will production be generally decentralized? How much energy storage will be required and in what form?

Not shown here but also available from IIASA is a breakdown of final energy use by sector, with the "business as usual" scenario dominated by industrial demand and the environmentally constrained one by a mix of transport and industrial demand growth.

The transitional and the developing economies clearly provide most of the growth in energy use, which again raises the question of supply, resources and equity. Figure 2.2 suggests that a large proportion of energy (and this must include the developing economies) will come from fossil fuels and from nuclear power, while Figure 2.3 suggests that local resources – mainly renewable – will be dominant. The decisions that will govern these choices will be made during the next decade and are directly associated with the drivers discussed in the first section. The major influences are likely to be in market liberalization and the regional and national responses to it, and in technological development and uptake. Technology transfer and the proposed

Figure 2.5
Projection for World Primary Energy Use to 2100
under Environmental Constraints by Region

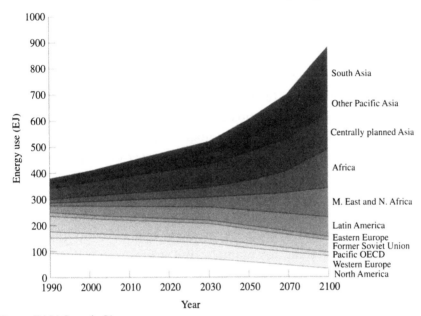

Source: IIASA Scenario C1

"Clean Development Mechanism" (CDM) under the Kyoto Protocol may prove crucial in this context.

Figure 2.5 shows that the really significant potential for reducing energy use is in North America – currently the area with the highest energy intensity per capita. The population growth of the developing regions and the very low current per capita energy use will ensure that there are no reductions there, but significant savings could be made in North America in improved transport efficiency – as in all sectors. Tighter building regulations and more efficient appliances alone could make a large impact.[14]

Projections for Carbon Emissions

An important driver of future energy use will be the responses to the Kyoto Protocol, both initially with regard to ratification and the realistic likelihood of signatory states reaching their targets, and then in the period "after

Kyoto." It has been pointed out that if climate change is real, then achieving Kyoto targets is only the first step on a much longer road, with cuts of closer to 60 percent in GHGs needed to stabilize their concentration in the atmosphere. This would entail not only major shifts in current energy use in the developed world, but also the use of non-fossil sources for the developing economies as shown in Figure 2.3.

Carbon emissions projections are highly variable, and many have been assessed under the work of the IPCC. However, only in the most aggressive environmentally driven scenarios do emissions drop or even stabilize. Nevertheless, development of technologies and mechanisms that takes place because of the "precautionary principle" over the next decade may have a significant impact on changing energy paths. In particular, technology transfer to the developing world, which is seen as one of the more promising opportunities for implementing proposals within a CDM framework, may find rapid acceptance in areas that have no entrenched electricity supply system, for example, and no institutionalized way of thinking.

Structural Changes

Changes in the structure of energy markets are already being seen, as liberalization takes hold, and hints of further changes are also visible. The former are mainly concerned with who wields the power and who is able to participate in energy markets, until now restricted to a favored few. It will take some time before the various markets, in their different stages of evolution and under different sets of rules, settle out to equilibrium and display the behavior that is expected by some observers. At the two extremes this may be an efficient, low-priced, responsive market, or a restrictive and monopolistic framework in which reliability is low and prices are high.[15]

Other structural changes may be more technical yet perhaps more subtle. The refinery industry in Europe and in North America is being forced to produce higher levels of low-sulfur fuel by legislators within the two regions.[16] This is slightly more energy-intensive than current operations, perhaps increasing carbon emissions, but is radically different in terms of hydrogen production. A current average of 0.5 percent w/w of hydrogen per barrel of oil is projected to increase to 1.2 percent w/w by 2010 to fulfill requirements, and this hydrogen must be either produced by the refineries from feedstocks they have, or bought in. At the same time, there is likely to be an increase in the requirement for alternative fuels, possibly hydrogen, under the ZEV mandate in California. If more hydrogen is required both in the refinery and in the fuel market, then there will be strong forces at play directing the future production of fuels.

Possible Outcomes

I believe that there is an Arabic saying that expresses a robust view of predictions: *"Kathaba al-munajimun walau sadaqu,"* which can be roughly translated to mean: *"All astrologers [and, by extension, forecasters] are lying, even if, by chance, they are later proved right."*[17] It would be unwise to consider the scenarios given above as anything other than that. They should be treated as ways of viewing the future so that plans for development of a particular path can be brought forward, or the possible consequences of another may be analyzed.

Under the scenarios, the continued use of oil as a fuel for the next 10 years is almost unchanged, as might be expected without a major intervention on the side either of the producers or consumers. After that, the scenarios rapidly diverge, with reductions in oil use as might be expected under a scenario that is driven by environmental pressures, and increases under one that is primarily determined by the growth of the world economy. However, to understand the potential for the scenarios to happen requires further investigation of the alternative energy sources that may come into play.

Alternative Energy Sources

This chapter is intended to analyze the potential for alternative energy sources. As has been shown in the previous section, that potential is large but varies dramatically under different conditions. More useful analysis can be made of the pathways that may be taken to reach some of those outcomes, and of the immediate pressures and opportunities that exist in the alternative energy arena. In addition, rather than restricting an analysis to alternative energy sources, alternative energy carriers must also be considered. These are likely to bridge the different energy sources and technologies, and in an integrated system could both enhance efficiency and reduce pollution.

Renewable Sources

Conventional energy sources have come to mean those that are primarily fossil-based, although some one billion people still rely almost exclusively on animal wastes and wood as energy sources, neither of which is fossil in nature. Alternative energy sources include this dung and wood under the definition of biomass, but also include many other resources. Solar power, both thermal and photovoltaic, wind and wave power, tidal streams and geothermal energy can all be counted as alternative and generally as renewable sources. Some of these, such as different forms of biomass, can be used as fuels and burned to provide heat or motive power; others produce electricity directly for use or for feeding into a local grid system.

However, alternative technologies for energy conversion and use should also be considered, along with alternative energy carriers, as these may have a significant impact on the spread of both renewable energy and conventional fuels.

This chapter does not go into exhaustive detail on renewable energies; this area is being covered by another chapter in this volume.[18] A brief summary of renewable energy sources is given here, followed by a more detailed outline of some alternative technologies.

Biomass

Biomass has tremendous potential to supply future energy requirements, with many different possible sources and conversion methods. Energy crops such as short-rotation forestry (willow, for example), starchy plants (such as tubers) and sugary plants (perhaps sugar cane) can all be converted into energy. The former are best burned directly to provide heat or to drive turbines to produce electricity, though they can be gasified or pyrolized in low-oxygen processes to produce syngas or liquid fuels. These fuels can again be used for combustion for heat and power, or in transport applications. Other forms of biomass include the aforementioned animal (and human) wastes, which can equally be burned directly or processed into gases or liquid fuels. While some care must be taken that energy crops are not grown as a substitute for necessary food crops, it is often possible to use the residue of the latter as an energy source or to use crops from areas that are unsuitable for growing food.

Biomass is particularly attractive as an energy source in many of the wetter tropical regions, where growing seasons are long and growth rates high. Under careful management, crops such as sugar cane can provide both food and energy from the unused residues, though there may be a requirement for further energy inputs in the form of fertilizers to the soil as future crops grow.

Biomass energy is suited to distributed uses, as it is generally diffuse and transporting it significant distances may use more energy than can be derived from the original source. This also points to more significant markets in developing economies.

Wind Power

Wind power has made substantial progress over the past 15 years: annual energy output per turbine has increased by two orders of magnitude and 3–5 MW machines are under development. Variable speed operation is now possible thanks to power electronics developments, increasing the efficiency and the availability of turbines. However, there are some public perception problems with wind turbines, brought about by their size and the frequent requirement to site them in areas of scenic beauty where wind resources are great. Wind turbines do have some effect on bird life though it is small in comparison with overhead power lines, and their visual and aural intrusion characteristics must also be considered in the light of other technologies. In many areas where wind turbines have been installed, the local community has become more supportive after the event.

The world market for wind power is currently growing at about 35 percent per annum, as government policies encourage the installation of an increasingly more economic renewable technology. A "typical" wind farm can produce electricity at costs of about US$0.04/kWh, though the economics are highly site-specific, and as production continues to increase the costs will fall further, predicted to be in the region of 50 percent over the next two decades.

Current markets are largely in the developed world, but future markets in the developing economies are expected to be considerable. Offshore wind farms are now being developed in Western Europe to take advantage of the generally greater offshore wind speeds, while avoiding extensive land-based development.

Solar Power (PV and Thermal)

Many different types of solar power converter exist, including solar thermal concentrators, based on the concentration of light through arrays of mirrors, and silicon-based semiconductor devices that produce electricity directly from incident light energy. Comparatively little development work is being carried out on thermal systems, though there are successful trial installations in desert areas such as the Mojave. Although strong potential may exist for future cost reduction, it is still an expensive technology.

Equally expensive but more flexible are the photovoltaic arrays, built from a wide variety of different materials and with different characteristics. Typically, highly efficient but very expensive arrays can be constructed, or cheaper, flexible films used that have lower conversion efficiencies. Trade-offs must be made when the merits of either are being considered.

Existing systems are not cost-competitive with other forms of generation except in areas where there is no other form of generation. They require low maintenance and are highly reliable and can therefore be a sound invest-ment, particularly if there is no grid connection, despite the fact that they produce electricity at US$0.25–0.50/kWh.

Despite the high costs, photovoltaics are often considered to be a suitable technology for developing countries, where the cost of increasing grid cov-erage and the comparatively high insolation in many regions make them directly applicable to the local situation. Rapid future growth in developing world markets is expected.

In the developed economies there is also significant potential, largely in the arena of building-integrated photovoltaics, where the material can be used both as a cladding (displacing some costs) and as an energy producer. This could result in significant decentralized electricity production in some of the sunnier areas of developed nations.

New Technologies

Fuel Cells

Five different basic types of fuel cell exist, each with its own operating char-acteristics and potentially suitable applications. Low-temperature fuel cells, operating below 100°C, include the solid polymer (SPFC) and alkaline

(AFC) types, which prefer to operate on comparatively pure hydrogen and which may be most suitable for small-scale power generation (perhaps with some cogeneration of low-grade heat for space heating) and transport uses. The SPFC system is particularly suited to the latter as it has a solid electrolyte and high power density. High-temperature systems include the phosphoric acid cell (PAFC), operating at 200°C, and the molten carbonate (MCFC) and solid oxide (SOFC) systems, which operate at 650°C and 1000°C respectively. The latter two types are suitable for combination with gas turbines for high electrical efficiency of up to 70 percent, or industrial steam or process heating requirements. They are also capable of operating directly on fuels such as natural gas, converting them to hydrogen internally and obviating the need for additional fuel-processing systems. A full explanation of how the different fuel cell types work is inappropriate here, but can be found in many references,[19] along with a detailed discussion of relevant markets and applications.

Fuel cells still require research and, more especially, development. However, they have the potential to be the key "pivot" application that induces a transfer to the so-called "hydrogen economy." They operate at high efficiencies and produce negligible or zero pollutants, the latter if they are operated using only hydrogen rather than a hydrogen-rich gas derived from hydrocarbon fuels. They are also quiet, using an electrochemical process rather than thermal combustion and mechanically driven generators to produce electricity. Many prototype and pilot fuel cell systems exist, with the nearest to true "commercial" sales being a PAFC unit from ONSI of the USA. This 200 kW PC25 system currently costs US$3,750/kW or US$750,000 in total before installation. The system is provided in containerized form, requiring prepared ground and hook-ups for fuel, electricity and water, and includes both fuel processing and electrical power conditioning. At present, however, only the AFC is cheaper, at approximately US$3,000/kW for a system from ZeTek Power of the UK that has as yet far fewer credentials than the PC25, which has approximately 200 installed units worldwide.

Fuel cell systems are still in a developmental stage and require continued research and development to enable costs to be brought down and reliability to be proven. There are, however, niche markets in which they can be competitive early on, with an aim to building up production volumes and reducing costs through economies of mass-manufacture and of learning.

These markets include ZEVs such as in California, remote power applications where batteries may otherwise have been used, and portable applications such as laptop computers and mobile phones.

The inherent modularity of fuel cells is of key importance to their potential. They are not subject themselves to increased efficiency with scale, as all thermal devices are, and so can provide electricity at a rate of 50–60 percent efficiency even at scales of several kilowatts, which is impossible with other generating means. This, in turn, means that they can be managed as decentralized systems and that capacity can be developed incrementally as appropriate, rendering them in principle ideal for a wide range of projects, including community microgrid power in liberalized energy markets. In the longer term, once potential problems such as fuel provision, maintenance and reliability have been solved, they could also be used for small community projects in the developing world. The importance of the modular system demonstrates itself here as in other markets, in that comparatively low capital risk is associated with a small project, though the benefits in terms of electrical efficiency will be equal to a much larger investment, such as a 300 MW gas turbine, with a longer lead time and longer payback period.

Fuel Cell Fuels

While low-temperature fuel cells must run on comparatively pure hydrogen, as some elements (such as sulfur) act as a poison to the fuel cell catalyst, hydrogen can be produced from primary resources such as natural gas or biomass. The conversion step is important, as it is where a large proportion of the emissions from the fuel cell fuel cycle are generated, but analyses have suggested that fuel cells are generally considerably cleaner than comparable conventional systems.[20] In fact, hydrogen can be produced from any hydrocarbon compound and also through electrolysis of water, though the efficiency of the process and the emissions will depend substantially on the source of primary energy. The fuel cell can thus be considered to be "fuel-flexible" in some ways, giving it the opportunity to operate in many different environments. However, the requirement for hydrogen, at least in the low-temperature versions, may have a wide influence on future fuel use and on energy patterns that will be discussed in the final section.

Microturbines

Another growth technology of the 1990s has been the microturbine, derived from automotive turbocharger technology or from aerospace and defense uses and available in units of between 30 and 100 kW. Like the fuel cell, the microturbine offers comparatively good efficiency for a small unit, though less than the fuel cell, at about 30 percent. However, it is currently much cheaper and more advanced, with current costs of about US$1000/kW and forecasts of US$500/kW in the near future. They are also flexible with regard to fuel input but are frequently designed to operate on natural gas, and have low emissions and low noise in comparison with conventional reciprocating engines of similar size. Their use in decentralized generation applications is already under evaluation, and it is possible to couple them with small fuel cells to increase the electrical generating efficiency of the package.

The use of microturbines is again an important consideration in liberalized energy markets and potentially in off-grid applications, perhaps in developing countries. Unlike the fuel cell, however, microturbines do not require hydrogen and thus do not have the same potential for change regarding energy carriers that may be used in the future.

Energy Carriers

Energy carriers may be primary energy resources, such as oil, but are more generally a refined form of the original, such as petroleum. Throughout history, there has been increasing sophistication in the energy carriers available, moving from wood and dung to coal, oil, gas and refined or generated products such as naphtha and electricity. The technology changes discussed early on in the chapter have gone hand in hand with the development of energy carriers, and new markets have emerged in conjunction with them.

However, it is not clear what future energy carriers will be used, or even whether the range of carriers will tend to increase or decrease. An interesting point to bear in mind is shown in Figure 2.6, derived from work done by Cesare Marchetti and others at IIASA.[21] This shows the continuous reduction in the carbon content of energy carriers per unit energy over the past 150 years, an intriguingly consistent process. There are several possible reasons that may underlie this trend, one of which is the relatively higher

Figure 2.6
The Decreasing Carbon:Hydrogen Ratio
in World Energy Use over Time

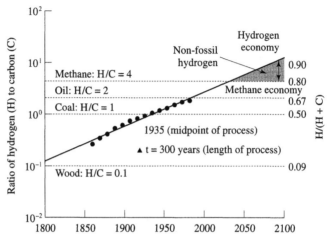

Source: Grübler et al. 1999

energy content of hydrogen per mass than carbon – as the hydrogen:carbon ratio rises, there is more energy available in the same mass of fuel. Hydrogen is also cleaner since burning natural gas produces lower emissions than burning coal, for example. As technologies have developed to handle these lighter and more generally difficult fuels, they have increasingly entered into common use because of these inherent emissions and energy advantages.

The future may see the continued development of what is referred to in Figure 2.6 as "the hydrogen economy." However, while the proportion of carbon in the energy mix may continue to decrease, there are energy carriers besides hydrogen that can fulfill many requirements.

Electricity

It has been suggested that the most likely long-term energy carrier in the future is electricity, in particular because many renewable energy sources produce it directly. However, electricity is a physical phenomenon rather than a fuel and cannot be easily stored, so it is not a useful form of energy for some applications. This leads to some difficulties with renewably generated electricity, as the latter is often produced from intermittent resources.

The sun only shines during the day and the wind blows at different speeds. So, even if the peak capacity of the renewable resource is able to meet the peak load, there may be a mismatch in timing. Some renewables, such as biomass, do not suffer this problem and may be able to help in backing-up the system, but without energy storage significant problems exist.

Energy storage can take many forms: pumped hydro installations, battery banks, compressed air in caverns, ultra-capacitors and flywheels are a few of the options. Each has advantages and drawbacks, which are discussed in more detail in other places and would be distracting to analyze here.[22] Gaining in interest, however, is the potential use of hydrogen as a form of energy storage.

Hydrogen

Hydrogen has been described above as a potential energy carrier for the future, and indeed of the past and present. Approximately 50 percent of the composition of town gas, which was made from coal and coke before the widespread introduction of natural gas, was hydrogen. The space industry has long used hydrogen as its fuel of choice. Although it can be difficult to handle, it is versatile and can be used both as stationary energy storage and as a transport fuel, particularly if the fuel cell can be introduced economically into the world's energy and transport infrastructures.

Hydrogen is easily and efficiently produced from water through electrolysis, and can then be stored as a compressed gas at a wide range of pressures depending on the application. It can then be recombined with oxygen in a fuel cell to produce electricity, water and some heat, or burnt in a slightly modified internal combustion engine. The former process is emissions free, while the latter may have traces of NO_x and hydrocarbon emissions, in part from lubricating oil. In either case, the hydrogen can be used either for stationary or transport applications, signaling the potential for future integrated zero emissions energy structures.

While hydrogen can be easily produced, the capital and operations cost of all the ancillary equipment, which includes compressors, storage tanks and even dispensers for vehicles, is very important. The hydrogen industry is mature but reliant on large industrial gas companies providing specific high-purity hydrogen to customers with well-defined needs, and has little experience of the sort of mass-manufacturing required to reduce costs and encourage the global dissemination of its products. Safety and standards procedures, as with all fuels, need to be designed and implemented. However, renewable production of electricity is partly handicapped by its intermittent nature, particularly in liberalized electricity markets designed on the basis of past conceptions of large-scale grid developments. If electricity produced from renewable energy sources can be stored and then used on demand, it will be more competitive, and hydrogen storage enables this to be achieved while maintaining the "green image" of the product. The economic viability of a renewable electricity project may also be enhanced by the addition of energy storage in the form of hydrogen, as not only will the dispatch of power be controllable to match instantaneous loads, but there will also be an opportunity for arbitrage between high and low electricity prices in the local bidding mechanism. Not only does the risk of not being able to supply go down, but the potential reward gained by price arbitrage makes the project more attractive financially.

Conversely, a fuel cell system could also benefit from being part of an integrated package of renewable energy sources. The price of hydrogen is very location specific,[23] and the potential of a guaranteed supply of pure hydrogen from an electrolyzer could enhance the economics of a stationary system. Prototypes and demonstrations have been and are being constructed, particularly in areas where the price of conventional energy is high, renewable resources exist and the environment is a specific factor, such as in Alaska and Northern Canada.[24] Interest is also surfacing in areas where grid congestion may become a problem, such as in many major conurbations. If increasing electricity demand within the city cannot be met through expansion of the national transmission infrastructure to that point, or some parts of the local distribution network are overloaded, then local energy storage and dispatch may provide a solution. Analysis into the potential for installing electrolysis units and fuel cells near electricity substations is beginning in some areas.

An interesting and potentially important benefit is the fact that hydrogen produced in these areas could be used not only for power support but also as a transportation fuel. While the infrastructure to supply hydrogen vehicles is under construction, there may be opportunities for supplying fuel to local depot-based fleets or to private vehicles, ensuring another revenue stream and perhaps also enhancing the uptake of vehicles requiring hydrogen.

The Developed and Developing Worlds

In understanding the future of world energy patterns, the relationship between the developed and developing worlds requires clarification. While most energy use is currently in the former, there is no doubt that this picture will change rapidly and radically, as suggested in the earlier scenarios, and it is the developing world that will become most relevant. This process is driven both by increasing populations and by industrialization and improvements in standard of living. Supplying all of this energy by traditional means will entail substantial developments in supply infrastructures, while the adoption of more non-traditional forms of energy will also need significant invest-ment. There may be opportunities for oil in emerging environmentally benign energy paradigms, but only under strict controls.

Developed World Energy Changes

While the transition within the developed world will take considerable time and has to overcome massive inertia, there are signs of some movement toward different paradigms of energy use. Liberalized energy markets are allowing innovation in some sectors, and the most forward-looking compa-nies are already finding ways of managing energy without selling kilowatt-hours of electricity or therms of gas. TXU Europe has introduced a "comfort" tariff for certain markets, under which the customer is guaran-teed a fixed level of comfort (in this case defined by the temperature within a dwelling) for a fixed payment. This leaves TXU free to manage the energy flows and invest in insulation and other measures if necessary, with the intention of maximizing efficiency and minimizing cost.

Carbon taxes have already been introduced in some countries: Italy, Norway and Sweden each have them in some form and many other countries are moving toward them. The achievement of Kyoto targets is under detailed discussion and some countries have put forward strategies and policies to begin meeting their objectives. At the same time, the influence of other driv-ers is becoming clearer, with urban pollution problems inspiring car-free days in European cities.

Corporate thinking also appears to be changing, with giant companies such as Asea Brown Boveri (ABB) divesting itself of large-scale generating plant interests and concentrating instead on the decentralized and renewable electricity sectors. British Petroleum (BP) is now the largest solar photo-voltaic cell producer, though Shell also has significant capacity in this area. Environmental qualifications are now highly valued in the financial sector, where corporate environmental responsibility is increasingly viewed as essential for continued business and profits; shareholders with major companies are querying strongly the environmental records of their companies.

Despite this apparent change in thinking, the inertia within the developed world, as well as the cost and complexity of developing a replacement infrastructure, will ensure that changes in the energy sector are gradual. They are likely to be increasingly influenced by environmental decision-making, but for oil it may be that security of supply issues weigh more heavily. In September 2000 concern in Europe over high fuel prices led to blockades of refineries, shortages and panic food buying. This resulted in briefly increasing the awareness of the local populations that they were dependent on, and hence vulnerable to, changes in the price and availability of something generally taken for granted. The ramifications of this situation will be unclear for some time following the initial disturbances, but it may result in policy thinking and changes that have a much broader effect on oil pricing and use.

Developing World Energy Futures

In the developing world things are far more open. Energy is desperately needed and the infrastructure generally poor, inadequate or non-existent. However, populations and expectations are both rising, and investment in the energy sector is high. There seems to be some awareness of a need to design energy systems appropriately for the regions in which they will operate, rather than transferring technology directly from a different country, though this is not universally true and investment decisions are frequently skewed by available overseas funding or aid.

It is to be hoped, however, that the mechanisms of the Kyoto Protocol may be used flexibly to enable technology transfer and appropriate use of resources to be developed for many countries, particularly under CDM. While the details behind this mechanism are not decided, the expectation is that low-carbon technology projects financed by the so-called Annex I countries (those subject to the Kyoto Protocol) can be installed in less developed countries (LDCs). Where they are shown to have offset carbon emissions that would otherwise have taken place, the benefit can be "credited" to the financing country. These technologies are more likely to involve renewable electricity production, clean transport solutions, biomass, hydrogen and fuel cells than oil projects, though carbon sequestration may be included in the portfolio of eligible technologies.

There is strong potential for integrated solutions to be developed: remote areas where there is no grid but solar or wind resources could be the first to have new technologies installed. Once fuel cell systems have been thoroughly proven, they should be low-maintenance, having few moving parts. This is potentially very valuable in remote areas. In addition, the transport problems of the LDCs are compounded in their mega-cities. Cairo, Mexico City, Delhi and São Paulo have some of the most polluted urban air, and are currently the focus of Global Environment Facility (GEF) projects investigating the potential of hydrogen fuel cell buses – both to improve the air quality within the cities and to reduce CO_2 emissions from the sector.

The use of fuel cells in both stationary and transport applications in LDCs is often not bound by the infrastructure constraints apparent in many of the developed economies. The possibility of sharp changes in energy structure thus becomes greater if a successful technology can be introduced into a particular sector, for example. Predicting change, growth and future resource needs is thus much less certain for LDCs than it is for the more entrenched developed economies.

Possible Effects on Oil

Following simple technology substitution patterns, it is easy to see that petroleum products could not be phased out for a very long time, even if there was a strong desire for some countries to do so. Even the most optimistic projections show that the internal combustion engine will continue to dominate transport for some considerable time – 20 years at a minimum,[25] and that the penetration of fuel cell vehicles, for example, cannot happen quickly. The LDCs, while they arguably have the opportunity to switch to fuel cells more easily as they have little entrenched fueling infrastructure, frequently suffer from a lack of support technology as regards repairs and maintenance. Fuel cell buses may be an important opportunity with respect to more advanced urban regions such as those discussed above, but even they will take time to penetrate the market. The dominant fuel will continue to be petroleum-based for some time to come.

What Does the Future Hold?

The future, as the past, "is a foreign country. They do things differently there."[26] However, it is possible to make long-term guesses based on short-term observations, and it appears that there are many drivers that are now aligned toward the uptake of alternative energy technologies in a considerably more serious way than has been true in the past. In the not very distant future it is possible to see integrated energy solutions enabling the best use of existing energy sources, and the introduction of new ones. In the long term it is quite feasible that the energy structure will have changed radically.

Moving to the Long Term: 2100

Current research and technology development, combined with the apparent impact of market forces, suggest that long-term visions of energy may be decentralized ones. Technologies such as fuel cells, solar power, wind power and biomass are often as efficient at small scale as they are at large scale, and can thus be installed in appropriate areas to fulfill, for example, electricity requirements at small capital costs in comparison with large-scale plant and extension to the grid network. These technologies tend to be oil-independent, and in some areas will displace technologies that previously operated using petroleum products.

Requirements for low or zero urban emissions will probably force the introduction of the fuel cell vehicle, and hydrogen storage technology, currently an area of increasingly heavy investment, should have developed to allow hydrogen to be the fuel of choice for all fuel cell vehicles. The preferred energy carriers will thus be those that have zero emissions in use and can be generated from renewable resources – electricity and hydrogen. Fossil fuels in general will play a small part in the long-term future.

Pathways from 2000 to 2100

Before 2100, however, a considerable amount of oil will be used in many sectors, although most of it will be used in transportation. Demand will almost certainly continue to rise for the next decade as the appetite of the developing world increases. The transition away from oil will start gradually and will almost certainly begin with fleet transport and with specific remote areas for stationary power. Hydrogen is already in use as a fleet-vehicle fuel in Hamburg, Munich and Palm Desert, with demonstrations in Chicago and Vancouver already over and waiting for a next stage. By 2003 ten European cities will have hydrogen-fueled demonstration bus fleets, and GEF projects could be under way in some of the mega-cities mentioned above. High-profile demonstrations of actual capability should allow both public perception to be enhanced and industry opportunity to be increased. California's retention of the ZEV mandate may prove crucial in this context, first as a large market in itself but also because a consortium of north-east states, including New York and Massachusetts, may adopt Californian legislation without alteration.

Following these fleet vehicles, it is likely that private fuel cell vehicles will become increasingly important, although the speed of introduction depends crucially on the development of the infrastructure for refueling them. Predictions of timescale cannot be clearly made. Nevertheless, in some countries there is a compelling need for new solutions, and private two- or three-wheeled vehicles produce disproportionately large amounts of pollution. The use of fuel cell scooters in both Taiwan and India is under serious consideration and could link well with locally produced hydrogen.[27]

At the same time, some stand-alone renewable systems and grid-connected renewable systems with a requirement for energy storage will be integrated with fuel cells and hydrogen storage, able to supply both transport fuel and power as required. The changing make-up of the refinery and its increased hydrogen production capacity may also enable it to sell hydrogen as a fuel into a local area, while this is more profitable than using its production on-site.

Gradually, there will be penetration of fuel cell technology using local resources into the LDC energy systems, and oil products will become less valuable and decline in use.

Thoughts about the Effects on Oil

Sheikh Yamani suggested in summer 2000 that the oil age was about to be supplanted by another age, through the introduction of the fuel cell.[28] While it is likely that he is right, it will not be until some time far in the future that full replacement takes place. The change, however, is particularly complex in nature, as it has elements of positive feedback. The formation of Shell

Hydrogen in early 1999, for example, may turn out to have speeded up the introduction of hydrogen and fuel cells to the market by raising the profile of the technologies and encouraging investment. The single-minded determination with which DaimlerChrysler has pursued fuel cell vehicle technology has almost certainly helped to drive the technology forward for the same reasons. The profit cycles of corporations thus can be strongly influenced by their own levels of investment in technology, and the uptake of alternative energy sources influenced accordingly.

The future for oil is certainly bright for the next 10 years, and probably for much longer, though technology and policy changes will have a major influence on trends in the latter part of the decade. It is probably prudent for both companies and nations that rely on oil for a major portion of their revenue to consider diversification into alternative energy sources, perhaps into a basket of renewable generating technologies, hydrogen and fuel cells. These will certainly become important at some point in the near future, and that point may be closer than can be accurately predicted.

3

Renewable Energy:
Now a Realistic Challenge to Oil

Timothy E. Lipman and Daniel M. Kammen

Conventional energy sources based on oil, coal and natural gas have been highly effective drivers of economic progress but at the same time damaging to the environment and to human health. Furthermore, they tend to be cyclical in nature, due to the effects of oligopoly on production, distribution and other factors. These traditional fossil fuel-based energy sources are facing increasing pressure on a host of environmental fronts, with perhaps the most serious challenge confronting the future use of coal, given the Kyoto Protocol greenhouse gas (GHG) reduction targets. It is now clear that, barring radical carbon sequestration efforts, any effort to maintain atmospheric levels of CO_2 below even 550 ppm cannot be based fundamentally on an oil and coal-powered global economy.[1]

In contrast, renewable energy systems are based on a paradigm of small-scale decentralization that addresses numerous problems regarding many electricity distribution, cogeneration (combined heat and power), environmental and capital costs. Instead of traditional on-site construction of centralized power plants, renewable systems based on photovoltaic (PV) arrays, windmills and biomass or renewable hydrogen-based fuel cells and microturbines are capable of being mass-produced at low cost and tailored to meet specific energy loads and service conditions. These systems can have dramatically reduced as well as widely dispersed environmental impacts, rather than larger, more centralized impacts that in some cases seriously contribute to air pollution, acid rain and global climate change.

The "Coming Age" of Renewable Energy is Here

The potential of renewable energy options has been apparent for some time, as noted in several seminal writings on the topic.[2] However, a new phenomenon is that the costs of solar and wind power systems have dropped substantially over the past 30 years and continue to decline, while prices of oil and gas have continued to fluctuate. In fact, one could argue that fossil fuel and renewable energy prices are heading in opposite directions, as more and more types of social and environmental costs are taken into account. Furthermore, the economic and policy mechanisms needed to support sustainable markets for renewable energy systems have also rapidly evolved. It is now becoming clear that future growth in the energy sector will primarily be in the sphere of renewable resources and to some extent in natural gas-based systems, but not in conventional oil and coal sources. Financial markets are awakening to the future growth potential of renewable and other new energy technologies, and this development probably heralds the economic reality of truly competitive renewable energy systems.

Recent Progress in Renewable Energy System Cost and Performance

Both wind and solar power have become much cheaper and more productive since their initial breakthrough following the first oil "price shock" of the 1970s. The capital cost to install wind energy systems has declined from about US$2,500 per kW in the mid-1980s to about US$1,000 per kW in the mid-1990s.[3] The American Wind Energy Association (AWEA) estimates that the current average cost of wind energy systems ranges from 4.0 to 6.0 cents per kWh, and that the cost is falling by about 15 percent with each doubling of installed capacity.[4] Installed capacity has doubled three times during the 1990s and wind energy now costs about one-fifth as much as it did in the mid-1980s. Advances in design and manufacturing, together with further economies of scale, are expected to bring the average cost of wind power down to 2.5 to 3.5 cents per kWh over the next ten years.[5] Wind turbine performance has also improved over time and is expected to continue to improve. The US Department of Energy (DOE) forecasts a 25–32 percent improvement from a 1996 baseline by 2010 regarding net energy produced per area, rising to 29–37 percent in 2020 and 31–40 percent in 2030.[6]

Solar energy technologies have also become significantly cheaper, but the capital costs of PV and solar thermal systems have not yet declined to the point where these systems can compete with the most economical fossil fuel and wind power systems. In Japan solar PV module prices have declined from ¥26,120 per watt in 1974, when the "Sunshine Project" was started, to ¥1200 per watt in 1985 and ¥670 per watt in 1995 (in constant Year 1985 ¥).[7] The DOE reports that from 1976 to 1994 PV modules have followed an 82 percent experience curve, which means an 18 percent reduction in cost with each doubling of production. Costs fell from over US$30 per watt in 1976 to well under US$10 per watt by 1994.[8] PV module costs currently range from approximately US$4 per watt for larger installations to about US$6 per watt for small residential or remote power systems.

In addition to the progress in cost reduction made by wind and PV systems, other renewable energy systems based on biomass, geothermal and solar thermal technologies are also experiencing cost reductions that are forecast to continue. Figure 3.1 presents forecasts made by the DOE for the capital costs of these technologies from 1997 to 2030.

Of course, capital costs are only one component of the total cost of generating electricity, which also includes fuel and operation and maintenance (O&M) costs. In general, renewable energy systems are characterized by low or no fuel costs, although O&M costs can be considerable. It is important to note, however, that O&M costs for all new technologies are generally high and can decrease rapidly as operational experience increases. Renewable energy systems, such as PVs, and other clean energy technologies (CETs), such as fuel cells, contain far fewer mechanically active parts than comparable combustion systems and therefore are likely to be less costly to maintain in the long term.

Furthermore, renewable energy systems have performed significantly better and this improved performance, combined with lower capital cost, has produced cheaper electricity. For example, wind energy costs are being reduced by the increases in efficiency of new, larger (megawatt class) wind turbines, and PV systems have experienced gradual but significant efficiency improvements. The best thin-film PV cells tested in laboratories in 1980 achieved efficiency levels of about 10 percent. This result was improved to about 13 percent by 1990 and over 17 percent in recent years for thin-film cells made from copper indium diselenide ($CuInSe_2$) or cadmium telluride (CdTe).[9] Figure 3.2 presents DOE projections for the average cost of electricity production from 1997 to 2030 by the same renewable energy technologies shown in Figure 3.1.

Given these likely average capital and system cost reductions, recent analyses have shown that additional generating capacity from wind and solar energy can be added at lower incremental costs relatively than additions of fossil fuel-based generation. The response to a recent study by the Center for Energy and Economic Development (CEED), a coal industry lobbying group, illustrates this point. The CEED estimated that a modest increase in non-hydropower renewable generating capacity in the US from 2 percent to 4 percent would entail incremental costs of US$52 billion. In contrast, the US National Renewable Energy Laboratory (NREL) estimated that the actual incremental costs of this increase would be just US$1.9 billion over 15 years or about US$100 million annually.[10] These incremental costs would be further offset by environmental and human health benefits. Furthermore, the NREL analysis showed that geothermal and wind energy could actually become more economic than coal during this 15-year period.

Another analysis along the same lines was conducted by the Renewable Energy Policy Project (REPP). This analysis shows that adding 3050 MW of

Figure 3.1
Capital Cost Forecasts for Renewable Energy Technologies

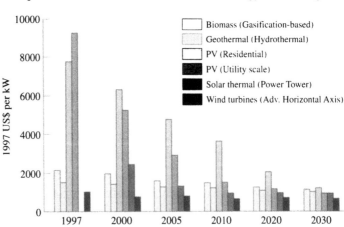

Source: USDOE 1997

Figure 3.2
Levelized Cost of Electricity Forecast for Renewable
Energy Technologies

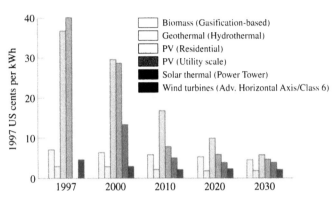

Source: USDOE 1997

Figure 3.3
Approximate Actual Electricity Costs (2000)

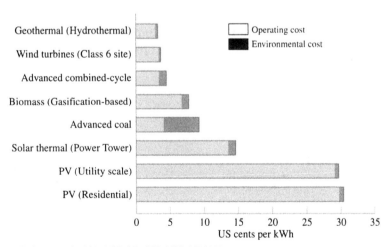

Sources: Ottinger et al. 1991; USDOE 1997; USDOE 2000

wind energy production in Texas over a 10-year period would entail only modest additional costs to residential customers. REPP estimates these additional costs to be about 75 cents per month for a household using 1,000 kWh per month, or about US$9 annually.[11]

The economic case for renewable energy resources and technologies looks even better when environmental costs are considered along with capital and operating costs. Figure 3.3 is an updated version of a chart originally published about 10 years ago.[12] The production cost figures have been updated on the assumption that the environmental costs have remained the same over this period. As shown in the figure, geothermal and wind can be competitive with modern combined-cycle power plants and, once approximate environmental costs are also included, geothermal, wind and biomass all have lower total costs than advanced coal-fired plants.

Renewable Energy: A Growth Sector and Business Opportunity

Renewable energy systems are experiencing significant growth at present and they are projected to experience further growth over the next few decades. Figure 3.4 indicates historical growth in global PV sales from 1990 to 1999, together with forecasted growth until 2010. Historically, the average growth in sales from 1990 to 1999 has been approximately 30 percent per year. The growth forecasted until 2010 is a 20 percent annual growth rate. Such growth will largely depend on continued PV panel cost reductions and

Figure 3.4
Global Growth in Photovoltaics Market

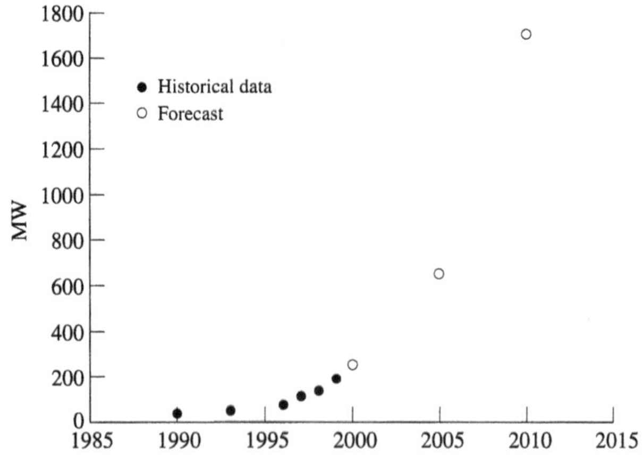

Source: BTM Consult APS 2000

Figure 3.5
Global Growth in Wind Power Market

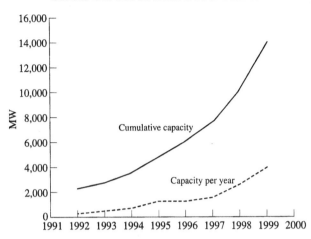

Source: Maycock 2000

Figure 3.6
OECD Electricity Mix (1970–2020)

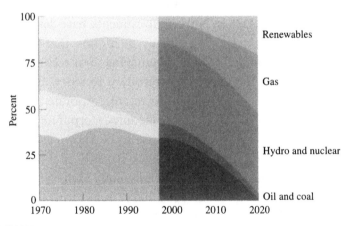

Source: Shell 2000

assumes that installed PV system costs will reach a range of US$3.00 to US$6.00 (in current dollars) by 2010, compared with US$8.00 to US$12.00 at present.[13] However, while the exact extent of future PV sales is uncertain, the trend is clear. Figure 3.5 presents an even more dramatic historical growth profile for wind power, with an average sales growth rate of 40 percent from 1994 to 1999.[14]

Shell Petroleum has made a high-profile projection of the future growth of renewable resources. As shown in Figure 3.6, Shell projects that renewable resources and technologies could constitute about 15 percent of the OECD's energy production by 2020, and that renewable resources and natural gas combined could account for about 50 percent of total production.[15]

As noted above, the remarkable difference between the context of renewable energy today and that of the past 30 years is that renewable and other clean energy technologies are actually now becoming economically competitive, and the incentive to develop them is no longer being influenced solely by environmental concerns. Merrill Lynch's Robin Batchelor recently referred to prospects for investing in companies developing clean energy resources, saying, "[t]his is not an ethical investment opportunity, it's a straightforward business opportunity."[16]

Mr. Batchelor also noted that the traditional energy sector had lacked appeal to investors in recent years because of heavy regulation, low growth and a tendency to be highly cyclical. He identified 300 companies worldwide that aim to develop wind, solar and wave power technologies and to advance capabilities in energy storage, conservation, and on-site power generation. Other investment firms, such as Nuveen Investments, have designed "renewable energy" and "fuel cell" investment portfolios that include large shareholdings of the companies that are pursuing these technologies.[17]

Further evidence of the impending transition to renewable energy and hydrogen-based systems can be found in recent corporate reorganization activities among the world's largest oil companies. Corporate giants such as Shell and BP Amoco, which would now like to be known as "energy companies" rather than "oil companies," have recently reorganized into a broader array of business units that include units exclusively focused on renewables and hydrogen. Such subsidiary units now include "Shell Renewables," "Shell Hydrogen," and "BP Solar."

Developing Country Case Studies and Lessons Learned

In developing nations renewable energy technologies are increasingly used to address energy shortages and to expand the range of services in both rural and urban areas. In Kenya over 120,000 small (20–100 Wp) solar PV systems have been commercially financed and installed in homes, battery charging stations and other small enterprises.[18] Meanwhile, a government program in Mexico has disseminated over 40,000 such systems. In the Inner Mongolia autonomous region of China over 130,000 portable windmills provide electricity to about one-third of the non-grid-connected households in the region.[19]

These case studies demonstrate that the combination of sound national and international policies and genuinely competitive markets – the so-called "level playing field" – can be used to generate sustainable markets for clean energy systems. They also demonstrate that renewable energy systems can penetrate markets in the developing world, even where resources are scarce, and that growth in the renewable energy sector need not be limited to applications in the developed world. Just as some developing countries bypass the construction of telephone wires by leaping directly to cellular-based systems, they may also avoid having to build large, centralized power plants and instead develop decentralized systems. This strategy can mitigate the environmental costs of electrification and can also reduce the need for the construction of large power grids by allowing for different topologies, based on "microgrids" and generation for local use.

Developing Renewables Along with Fossil Fuel Technologies: Synergies and Challenges

Despite their limited recent success, renewable energy sources have historically had a difficult time breaking into markets dominated by traditional, large-scale, fossil fuel-based systems. This is partly because renewable and other new energy technologies previously had high capital costs relative to more conventional systems and are only now being mass produced, but also because coal, oil and gas-powered systems have benefited from a range of subtle subsidies over the years. These subsidies include military expenditures to protect oil exploration and production interests overseas, the costs of railway construction that have enabled economical delivery of coal to power plants, and a wide range of smaller subsidies.

However, another limitation has been the intermittent nature of some renewable energy sources, like the wind and the sun. One solution to this last problem is to develop diversified systems that maximize the contribution of renewable energy sources, but that also use clean natural gas and/or biomass-based power generation to provide base-load power when the sun is not shining and the wind is not blowing. Using a range of different renewable energy technologies to provide energy for a region can also help to mitigate the intermittent nature that some of them exhibit. Even when there is no wind blowing, there may be strong solar insolation and vice versa.

In essence, however, renewable energy technologies face a situation confronting any new technology that attempts to dislodge an entrenched, "incumbent" technology. For many years we have been "locked-in" to an array of fossil fuel and nuclear-based technologies, and many of our secondary systems and networks have been designed and constructed to accommodate these. Just as vehicles powered by electricity will face an uphill battle to dislodge gasoline-fueled, internal combustion engine vehicles, solar, wind and biomass technologies will not easily supplant modern coal, oil and natural gas power plants.

This "technological lock-in" situation has several important implications. First, various types of feedstock and fuel delivery infrastructure have been developed over the years to support conventional energy sources and in some cases this infrastructure would require modifications to support renewable energy technologies. This would entail additional cost, tipping the table away from the new challengers. Second, the characteristics of conventional energy systems have come to define how we believe these energy systems should perform and new renewable energy technologies that perform differently than conventional technologies, such as intermittent operation, may raise doubts among potential system purchasers. Third, to the extent that new technologies are adopted, early adoptions will lead to improvements and cost reductions in the technologies that will benefit later users, but there is no market mechanism for early users to be compensated for their experimentation that later provides benefits to others. Since there is no compensatory mechanism, few are likely to be willing to gamble on producing and purchasing new technologies, and as a result the market is likely to under-supply experimentation.[20]

Hence, particularly in the absence of policy intervention, one may remain locked in to existing technologies, even if the benefits of switching to new ones outweigh the costs. There are numerous examples, however, of an entrenched or locked-in technology being first challenged and ultimately replaced by a competing technology. This process is generally enabled by a new wave of technology and it is sometimes achieved through a process of hybridization of the old and the new. Technological "leapfrogging" is another possibility, but this seems to occur relatively rarely. A prime example of the hybridization concept is the case of the competition between gas and steam powered generators, which dates back to the beginning of the century. From about 1910 to 1980 the success of steam turbines led to a case of technological lock-in and to the virtual abandonment of gas turbine research and development. However, partly with the aid of "spillover" effects from the use of gas turbines in aviation, the gas turbine was able to escape the lock-in to steam turbine technology. Initially, gas turbines were used as auxiliary devices to improve steam turbine performance and then they slowly became the main component of a hybridized, "combined-cycle" system. In recent years orders for thermal power stations based primarily on gas turbines have increased to more than 50 percent of the world market, up from just 15–18 percent in 1985.[21]

Furthermore, Brian Arthur has shown that increasing returns on the adoption of new technology or "positive feedbacks" can be critical to the outcomes of technological competition.[22] These increasing returns can take various forms: industrial learning, for example learning-by-doing in manufacturing, together with economies of scale, may lead to a decline in production cost. Networks of complementary products, once developed, may encourage future users. Information about product quality and reliability may reduce uncertainty and risk to future users. Adopted technology may turn out to be more compatible with other technologically interdependent systems. Where increasing returns are important, as in most technology markets, the success with which a challenger technology can capture these effects and enter the virtuous cycle of positive feedbacks may, in conjunction with chance historical events, determine whether or not the technology is ultimately successful.

Just as the hybridization between gas and steam turbines gave gas turbines a new foothold in the market, hybridization between gas and biomass-fueled power plants may allow biomass to eventually become a more prominent energy source. Hybridization of intermittent solar and wind power with other clean "base-load" systems may allow the proliferation of solar and wind technologies. With advances in energy storage such systems could ultimately become dominant. Once they are able to enter the market, through whatever means, these technologies can reap the benefits of the cycle brought by increasing returns to adoption, and this is already beginning to happen with several new types of renewable energy technologies.

Fuel Cells, Distributed Generation and Hydrogen as an Energy Carrier

Another interesting option is to use hydrogen as an energy carrier and a buffer for an energy system that is based largely on renewable energy sources. Hydrogen can be generated via electrolysis when the renewable systems are on-line, and the hydrogen can be stored for later use and then used in fuel cells to produce electricity when needed. The viability of such systems will depend on reductions in the capital costs of the renewable energy systems and fuel cells/electrolyzers, as well as the design of the system, but they are one promising option for eventually mitigating the intermittent nature of some renewable energy sources.

Fuel cells, and particularly proton exchange membrane (PEM) fuel cells for vehicle and stationary applications, may be a key enabling technology for the transition from a fossil-fuel based energy system to one that is dominated by renewable energy sources. Initially, fuel cells could run on hydrogen that is generated via steam reformation of natural gas, which is currently the most economical method.[23] Eventually, solar, wind, geothermal and hydro-power could generate electricity that produced hydrogen with electrolyzers, as well as biomass sources. The diverse array of feedstocks that can be used, along with the clean and efficient operation of the fuel cells themselves, suggest that fuel cells could contribute enormously to producing electricity and displacing petroleum in a variety of scales and settings.

Fuel cell vehicles and other clean vehicle technologies are promoted, but at present they face uncertain commercial prospects due to the high costs of fuel cell technology, which is not yet mass-produced or optimized for cost reduction. In addition, the forecasts of manufacturing costs during mature, high-volume production are conservative. Although fuel cell vehicles may cost US$4 thousand to US$5 thousand more than conventional vehicles, their lifecycle costs can be competitive due to their high efficiencies, lower maintenance requirements, and longer operational lives. Figure 3.7 shows the estimated lifecycle operating costs in cents per mile (cents/mi) of battery and fuel cell EVs compared with conventional vehicles and also including pollutant and GHG emission estimates and values for the California South Coast. As shown in the figure, even with significantly higher estimated capital costs, both battery EVs and direct-hydrogen fuel cell EVs (DHFCVs) can have comparable or even lower lifecycle costs than conventional internal

Figure 3.7
Vehicle Lifecycle and Emission Costs:
High Production Volume Central Case

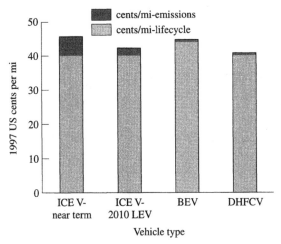

Source: Lipman 1999

combustion and low-emission conventional vehicles (LEVs), particularly once emission values for fuel upstream activities and vehicle operation in the Los Angeles area are considered.[24]

When fuel cell vehicles become a commercial reality, the potential implications will be enormous, going far beyond reductions in petroleum use, pollution and greenhouse gases. Electricity generated by fuel cell vehicles could be routed directly into homes or into the electric-power grid and used to provide peak power (or even base-load power) for non-transport uses. As an example, the current motor-vehicle fleet in the US (about 146 million vehicles) has a total power-generating capacity equivalent to about 14 terawatts. If used as generators, they could produce about 12 terawatts of electric power, assuming 85 percent efficiency. Amazingly, that amounts to approximately sixteen times the entire present stationary electric-generating capacity in the US.

Motor vehicles are driven an average of about one hour per day, so the generating capacity of the vehicle fleet is idle for approximately 95 percent of the time. A fleet of 100,000 fuel cell vehicles would be capable of producing about 2.4 gigawatts of power for the grid, assuming 25 kilowatts of net fuel-cell output of power per vehicle and 95 percent vehicle availability. Even if the vehicles were available for generating power only 50 percent of the time, those 100,000 vehicles could still contribute about 1.25 gigawatts or the equivalent of about 12 medium-sized power plants. Looking further ahead, if half of the vehicles in California's South Coast Air Basin were fuel cell powered (suppose by about 2020 or so), with each vehicle capable of supplying just 25 kW of power to the grid half of the time, the total generating capacity of these vehicles would be approximately equal to the present level of installed generating capacity in the entire state. (See Kempton and Letendre for similar calculations and further analysis of grid-connected battery EVs.)[25]

In addition to augmenting the capacity for electricity production, which is currently strained in the western USA and other regions, this type of scheme could also help to offset the higher initial costs of fuel cell vehicles compared to conventional vehicles. The manufacturing costs of fuel cell vehicles will decline rapidly as they move into mass production, but they may be somewhat costlier than conventional vehicles for some time. As noted above, fuel cell vehicles may cost a few thousand dollars more than conventional vehicles, in spite of mass production and some advances in technology. However, even if this proves to be the case, fuel cell vehicles can have comparable or even lower average operating costs to conventional vehicles due to their higher efficiencies, somewhat reduced maintenance costs and slightly longer assumed durability.[26] Since consumers tend to consider up-front vehicle costs more than they do the costs of long-term ownership, the initial capital cost difference may dissuade some consumers from purchasing these fuel cell vehicles. The possibility of recovering some of that additional cost with revenue from electricity sold to the grid, coupled with efforts to educate consumers about the real costs of owning and operating different types of vehicles, could mitigate this problem.

Thus, diverse systems using two or more types of renewable energy technologies with complementary capacity profiles, along with fuel cells that would run on natural gas or methanol and eventually on renewable sources, could provide the "stepping stones" from a fossil fuel-dominated system to one that is predominantly renewable and climatically stabilizing. This transition will not be simple, easy or without cost. However, the progress being made by solar, wind and biomass systems, together with the commercial promise of fuel cells, make this a likely scenario rather than an environmentalist's fantasy.

How to Level the Playing Field

As shown in Figure 3.3, renewable energy technologies tend to be characterized by relatively low environmental costs. In an ideal world this would aid them in competing with conventional technologies, but of course many of these environmental costs are not priced in the market. Only in certain areas and for certain pollutants do these environmental costs enter the picture. Clearly, further increased internalization of these costs would benefit the spread of renewables. The international effort to limit the growth of greenhouse emissions through the Kyoto Protocol may lead to some form of carbon-based taxation, and such a measure could prove to be an enormous boon to renewable energy industries. However, support for the Kyoto Protocol among industrialized countries remains relatively weak, particularly in the US, and any proposed carbon-based taxation scheme will surely face stiff opposition.[27]

Perhaps it is more likely that concern about particulate matter emission and formation from fossil-fuel power plants will lead to expensive mitigation efforts to reduce these effects, and this will help to tip the balance toward cleaner renewable systems. In a relatively controversial move, the US Environmental Protection Agency (EPA) has recently proposed new ozone and particulate matter (PM) standards that are even more stringent than the current standards that have remained unattained in some US urban areas. The EPA has justified these new regulations with estimates that show that new standards are necessary to provide increased protection against a wide range of potential health impacts. For example, the EPA estimates that even if Los Angeles County were to meet the existing PM standards, 400 to 1000 deaths per year would still occur as a result of exposure to very fine PM (under 2.5 microns in diameter) that is unregulated.[28] The combination of increased pressures to attain ozone and PM standards will further complicate the establishment of new fossil-fueled power plants in some areas of the US, particularly since they are now required to use processes that remove greenhouse gases from the atmosphere. This will indirectly but surely benefit renewable energy technologies, which do not typically face these difficulties in obtaining siting permits.

Public and Private Sector Investment Issues

A fundamental problem with any new technology is that it does not have the track record of performance that exists for older, more established systems. This simple fact is often cited by proponents of existing technologies in invalid arguments against technological change. New technologies and operational procedures present greater risks, but also greater opportunities for innovation and profit. A comparison of current costs for fossil fuel and renewable energy systems, shown in Figure 3.8, illustrates the greater range of costs for newer technologies.

Emerging energy systems were long seen as an area of risky investments, with the history of renewable energy systems seen as the primary illustration of that "fact." A variety of authors have shown that this perception is not only illusory but also largely a self-fulfilling prophecy.[29] The historian Richard Hirsh coined the term "technological stasis" to describe a process whereby companies – notably US utilities – chose to make only small incremental

Figure 3.8
Cost Comparisons of Mature and Emerging Energy
Generation Technologies

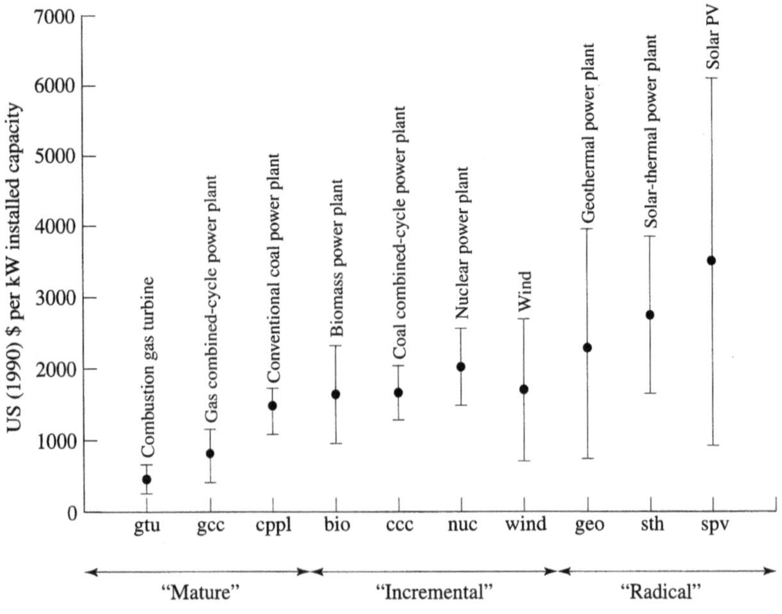

Source: Grübler et al. 1999

investments in innovation so as to slow overall technological progress. Larger gains, both to individual companies and to society, typically stem from carefully targeted but consistently pursued avenues of research, innovation and implementation.[30] Renewable energy systems now exhibit this feature of significant uncertainty, coupled with great promise and the potential for important innovations and profits.

Market Transformations

There are two principal reasons for government support of R&D to develop CETs. First, as noted above, conventional energy prices generally do not reflect the social cost of pollution. This provides a reason for subsidizing R&D on CETs as potential alternatives to polluting fossil fuels. Second, private firms are generally unable to appropriate all the benefits of their R&D investments. Consequently, the social rate of return for R&D exceeds available private returns, and firms therefore do not invest enough in R&D to maximize social welfare.[31] Thus, innovation "spin-off" among CET firms is a form of positive externality that justifies public R&D investment.

The conventional wisdom is that government should restrict its support to R&D and allow the private sector to commercialize new technologies. Failed CET commercialization subsidies (e.g. the US corn ethanol and synthetic fuel programs) bolster this view. Nonetheless, there are compelling arguments for public funding of Market Transformation Programs (MTPs) that subsidize demand for some CETs in order to help commercialize them. Furthermore, the argument that it may not be worthwhile for firms to invest in new technologies because of the spillover effects is generally false as well. As discussed above, early investment in new technologies in promising market sectors has proven to be the best strategy for firms interested in long-term and not simply short-term profitability (see Spence,[32] among others, for further discussion of this).

A principal motivation for considering MTPs is inherent in the production process itself. When a new technology is first introduced, it is invariably more expensive than established substitutes. There is, however, a clear tendency for the unit cost of manufactured goods to decline as a result of cumulative production experience. Cost reductions are typically very rapid at first but taper off as the industry matures. This relationship is called an "experience curve" when it accounts for all production costs, and it can be described by a progress ratio (PR) where unit costs fall by 100*(1–PR) per-cent with every doubling of cumulative production. Typical PR values range from 0.7 to 0.9 and are widely applicable to technologies such as toasters, microwave ovens, solar panels, windmills and essentially any good that can be manufactured in quantity.

For example, Figure 3.9 presents PRs for PVs, windmills and gas turbines. All three have initial PRs of approximately 0.8, which is a typical value observed for many products. However, after 1963 the gas turbine PR increased substantially, indicating an attenuation of experience effects.

Moreover, as with R&D, MTPs help to promote the use of CETs as alter-natives to polluting fossil-fuel technologies and thereby reduce the social costs of pollution. When politically possible, the best policy is to fully internalize pollution costs, for example through pollution taxes set at the marginal social cost of pollution or emissions permits set at the socially optimal pollution level that can be allowed. However, governments chron-ically fail to achieve this, providing another clear rationale to support MTPs.

When evaluating MTPs, it is essential to account for positive feedback between the response of demand and the effects of experience. An MTP increases the quantity produced in the first year and, due to the effects of

Figure 3.9
Progress Ratios for Photovoltaics, Windmills and Gas Turbines

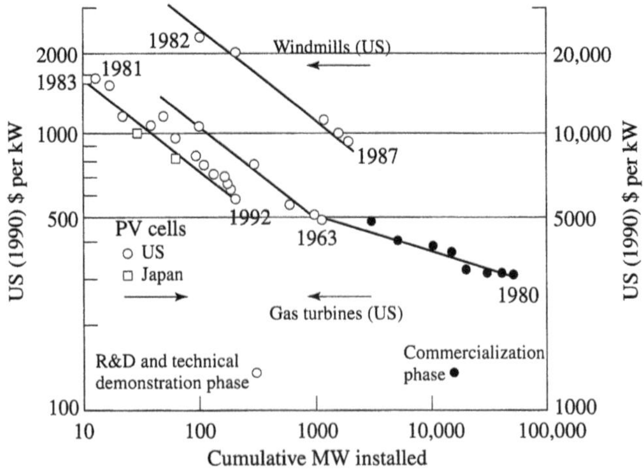

Source: IIASA/WEC 1995

experience, subsequent year unit costs are lower than they would have been without the additional production from the MTP. These lower costs, in turn, imply that the quantity demanded in subsequent years is higher. This "indirect demand effect," in turn, adds to cumulative production and further lowers unit costs in future years. This "virtuous cycle" process continues indefinitely, although it gradually dissipates once the MTP is discontinued. Duke and Kammen show that accounting for these indirect demand effects substantially raises the benefit–cost ratio (BCR) of typical MTPs.[33] Even without accounting for environmental benefits, their case studies of MTPs targeting PVs and efficient lighting show positive BCRs of 1.05 and 1.54, respectively. However, their assessment of federal subsidies for corn ethanol shows a BCR of approximately zero.

These results suggest a role for MTPs in national and international technology policies. However, the costs of poor program design, inefficient implementation or simply choosing the "wrong" technologies can easily outweigh cost reduction benefits. This suggests that MTPs be limited to emergent CETs with a steep industry experience curve, a high probability of major long-term market penetration once subsidies are removed, and a price elasticity of demand of approximately one or greater. The condition that they be clean technologies reduces the risk of poor MTP performance by adding the value of environmental benefits. The other conditions ensure a strong indirect demand. Finally, as with energy R&D policy,[34] public agencies should invest in a portfolio of new CETs in order to improve overall MTP performance through diversification.

Conclusions

In conclusion, we believe that the promise of renewable energy is rapidly becoming a reality. Both solar PV and wind energy systems are experiencing rapid sales growth, declining capital costs, declining electricity costs and continued improvements in performance. Because of these developments, market opportunity now exists both to innovate and to take advantage of emerging markets, with the additional assistance of governmental and popular sentiment.

While fossil fuels will remain in the fuel mix for the foreseeable future, current high petroleum costs, European gas shortages and protests, transient or not, illustrate the degree of social and political ill-will that energy insecurity can generate. The integration of renewable energy supplies and technologies can help to temper the cyclical nature of fossil-fuel markets and can give renewables a foothold from which they can continue to grow and compete. There are many opportunities for creative integration of renewables into energy production systems. These include combined fossil and biomass-fueled turbines, combinations of intermittent renewable systems and base-load conventional systems with complementary capacities. Fuel cells that will operate on hydrogen produced from natural gas and methanol, and eventually on hydrogen from solar, wind and biomass sources, provide another example. Strategies such as these, in conjunction with the development of off-grid renewable systems in remote areas, are likely to ensure continued sales growth for renewable and other CETs for many years to come.

At present, however, the level of investment in innovation for renewable and other CETs is too low. This is the case because of the imperfection of the market, which undervalues the social costs of energy production. The fact that firms typically cannot appropriate the full value of their R&D investments in innovation and that new technologies are always character-ized by uncertain performance and thus greater risk compared to their more well-developed rivals are factors too. These issues suggest a role for public sector involvement to develop markets for renewable energy technologies through various forms of market transformation programs.

Finally, current energy producers should perhaps consider the emerging renewable energy markets as an opportunity rather than a threat. Large energy companies are among those parties best positioned to capture new renewable energy markets because they have the capital needed to make forays into these markets. They also tend to have the most to lose if they do not invest and renewable energy technologies continue to flourish. Some degree of investment in renewable energy systems may appear a contradic-tion to energy producers, but it can be an appropriate hedging strategy if more stringent environmental regulation curtails the use of fossil fuels in the future.

The artful introduction and integration of renewable energy technologies into energy production systems, along with encouragement from the public sector where appropriate, can create a path that eventually leads to heavy reliance on renewable energy systems in the future. Such a future will be more environmentally and socially sustainable than the one we would achieve by following a more traditional path based on continued reliance on fossil fuels. This latter path in many ways implies higher risks to human and ecological health and well-being over time. It is a path that is increasingly difficult to justify given the performance that renewables are now starting to achieve.

4

The Future of Natural Gas in the Energy Market

Marian Radetzki

W hat is the future of natural gas in the energy market? The simple answer is: quite bright. This chapter will substantiate this assertion by reviewing, in turn, the recent history of consumption growth in major markets, the situation regarding production and resources in a short historical perspective, the prospects for gas in the world energy market, as seen by reputable forecasters like the International Energy Agency (IEA) and the US Energy Information Administration (EIA), the realism of these prospects (in view of recent developments that have made gas more competitive) and a future maverick, represented by the possible commercial breakthrough of gas transformation into middle distillates on a large scale.

Published statistics and analyses by public agencies and academic researchers will provide the backbone of the entire study, but dissenting conclusions will be indicated too. The chapter focuses on the future of natural gas and the time perspective adopted is some 20 years into the current century. For a proper understanding of things to come, it is essential to have a firm grasp of where we stand today and from where we have come. Hence, an important component of the analysis is a historical review, also covering a 20-year period. Gas volumes are expressed throughout in million tons of oil equivalent (MTOE), while prices are quoted in US dollars per million BTU ($/mmBTU).

The Role of Gas in World Energy Consumption

In 1999 gas satisfied just under a quarter of total primary energy needs in the world. This is a considerable rise from 18.4 percent in 1979 and 21.3 percent in 1989. Table 4.1 provides several pertinent facts about gas consumption. Since gas is very costly to transport, the major markets shown in the table are relatively self-sufficient, with little trade exchange from one to the other. They can therefore exhibit quite different characteristics, e.g. in terms of market growth, prices, overall gas penetration and gas uses.

West and Central Europe and the Former Soviet Union combined (Europe–FSU for short) represent the largest of these markets, with overall consumption representing more than 40 percent of the world total. Piped exports from the FSU cover a large part of the European gas deficit and supplies from Algeria provide a further source of gas to Europe. Gas penetration in the FSU (53 percent of total primary energy in 1999) is higher than in any other region, primarily as a result of an abundant wealth of

Table 4.1
World Gas Consumption

	MTOE			Percentage Increase		Percentage in 1999	
	1979	1989	1999	1979–1989	1989–1999	Share in Total Primary Energy	Self-Sufficiency
West and Central Europe	245	294	400	20.0	36.0	22.2	64
Former Soviet Union	307	570	483	85.7	−15.3	53.1	122
USA and Canada	567	546	620	−3.7	13.6	25.5	102
USA	*517*	*488*	*555*	*−5.6*	*13.7*	*25.2*	*88*
Asia Pacific	54	133	242	146.3	82.0	10.7	95
Japan	*20*	*43*	*67*	*115.0*	*55.8*	*13.2*	*0*
Middle East	32	89	158	178.1	77.5	41.5	107
Rest of World	65	97	161	49.2	66	21.5	137
World Total	**1270**	**1729**	**2064**	**36.1**	**19.4**	**24.2**	**102**

Sources: BP Amoco 1980, 1990, 2000

resources and efforts of the former communist regime, especially in the 1980s, to assure maximum use of this low-cost energy source. FSU gas consumption in the 1980s rose by an impressive 86 percent, but then declined in the 1990s as a result of the post-communist economic implosion. The share of gas in total primary energy nevertheless remained very high, because the use of other energy sources fell even more. Consumption in Europe grew only modestly in the 1980s, but the expansion gained speed in the following decade, not least because of the surge in gas usage for power generation. Penetration, at a relatively modest 22 percent of total primary energy usage in 1999, reflects the higher cost of supply compared to the FSU.

The North American market, defined for my purposes to comprise the USA and Canada, is completely dominated by the former country. It is also virtually isolated in that it receives only minute quantities from Mexico by pipe and from Trinidad as LNG. Gas penetration, at 25 percent, is close to the world average. Abundant low-cost coal supplies in both countries have constituted a constraint on wider gas usage. In the 1980s consumption declined in absolute terms to an important extent because of regulatory confusion, but the regime of regulation has since been cut and streamlined, resulting in a modest consumption in the 1990s.

In comparison with Europe–FSU and North America, Asia Pacific is a relatively small market, but with a record of phenomenal growth. Between 1979 and 1999 consumption has more than quadrupled, yet the penetration of gas was less than 11 percent in the latter year. Some countries with significant consumption, notably Australia, China, India, Indonesia, Malaysia, Pakistan and Thailand, mainly rely on domestic supplies. An important characteristic of this market is the large import volume of LNG, originating in the region, to Japan, Korea and Taiwan, three countries with hardly any domestic supplies. In 1999 such imports, in excess of 80 billion cubic meters, corresponded to more than one third of the market's total consumption. The high cost of transporting provides part of the explanation for the low gas penetration in the region.

The three regions covered so far account for some 85 percent of global consumption, and most analyses do not venture any further in the regional analysis. Table 4.1 proceeds one more step in singling out the Middle East as a separate regional market that is growing at even higher rates than Asia Pacific and fast approaching the Asia Pacific volumes. The penetration of gas in this market to more than 40 percent in the late 1990s has grown in strides as formerly flared gas is being put to use. The high prices of oil since the early 1970s have certainly speeded up the utilization of gas to free up more oil for highly profitable exports.

Data on end uses are not easily obtainable. In the OECD area as a whole, in the late 1990s, 25 percent of all gas was employed for power generation, 31 percent was absorbed for industrial usage, while households consumed 28 percent.[1] These shares vary among the OECD regions. Thus, in Japan almost 60 percent of all gas is devoted to power generation. In North America the lead sector is industry (32 percent), while in Europe households account for the greatest proportion of gas use (34 percent). Gas for power generation in the OECD has recorded by far the fastest increase among the sectors. From the early 1980s to the late 1990s it rose by 73 percent. Residential consumption expanded by 47 percent in the same period, while industrial usage increased by only 8 percent.

World Gas Resources and Production

It is clear that gas resources by themselves do not constitute a constraint on any plausible development of demand. Until a few decades ago there was not much separate exploration for gas. Gas was discovered as a byproduct during the exploration for oil, to the disappointment of the prospectors unless oil was discovered at the same time. Only since the 1960s have special efforts been made to establish gas reserves and the success has been substantial. As Table 4.2 indicates, the global 1999 gas reserves/production ratio is 62 years, much higher than the 41 years R/P ratio for oil. Given the recent emergence of specific searches for gas, the prospects for finding much more gas are highly favorable, especially since the science for gas exploration progresses quickly.

The table exhibits only three instances of production declines during decades and none of these were caused by constrained resources. The European production decline in the 1980s was primarily caused by Dutch attempts to restrict supply for monopolistic gains. The policy proved counter-productive for the Dutch, since it speeded up supply expansion from Algeria, the FSU and Norway, substantially reducing the Dutch market share. Indigenous European production, including that in the Netherlands, rose briskly in the following decade. The production decline in

Table 4.2
World Gas Production

	MTOE			Percentage Increase		Reserves/ Production (1999)
	1979	1989	1999	1979–1989	1989–1999	
West and Central Europe	201	197	254	–2.0	28.9	18
Former Soviet Union	336	668	591	98.8	–11.5	82
USA and Canada	577	537	633	–6.9	17.9	10
USA	*502*	*450*	*486*	*–10.4*	*8.0*	*9*
Asia Pacific	57	126	229	121.1	81.8	40
Middle East	32	92	169	187.5	83.7	>100
Rest of World	97	133	221	37.1	66.2	83
World Total	**1300**	**1753**	**2097**	**34.9**	**19.6**	**62**

Sources: BP Amoco 1980, 1990, 2000

the FSU in the 1990s was in direct response to the collapsing demand, caused by the macroeconomic implosion of the time. North America's production cuts in the 1980s, like the region's shrinking demand, are explained by an incomprehensible regulation of the gas sector in the US during that period. As the regulatory problems were overcome, North American production in the 1990s expanded at a pace exceeding demand.

The development of prices provides a better picture of the adequacy of resources and production to satisfy demand. Price stability over time would be strongly suggestive of a market in balance, with supply rising in parallel with demand. Rising prices over a longer period of time would be a warning signal of an inadequate supply situation, possibly caused by resource constraints. These relationships are valid despite the practice in Europe and Japan of tying the price of gas to the price of oil. If the price of gas, determined in this way, proves inadequate for stimulating new supplies, we will observe either market shortages or an upward creep in the average gas/oil price in new gas supply contracts over time. Neither phenomenon has been reported.

[113]

Figure 4.1
Prices for Natural Gas in Three Major Markets

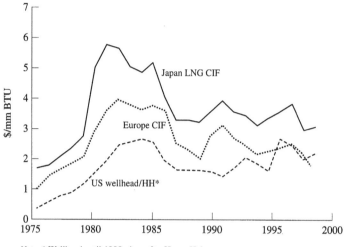

Note: * Wellhead until 1988; thereafter Henry Hub.

Source: BP Amoco, June 1990 and June 1999

Figure 4.1 and Table 4.3 depict prices in the three major markets over a period stretching back to 1975. The figure relies on annual data, hence the oil and gas price explosion of 2000 is not shown. Several noteworthy features should be observed. First, although all three price series are affected in the short run by the movements in oil prices (like the price spike 1980–1985), they reflect their specific market conditions and, by and large, vary independently of each other. Second, the substantial and continuous difference in price levels is mainly due to the differences in transport costs. Japan procures its supplies as LNG, with traditionally high costs for conversion and transport. Continental European prices are CIF West European border, with a substantial pipe transport component. The US price, in contrast, involves no transport cost at all until 1988. In the 1990s, however, the price is CIF Henry Hub in Louisiana, comprising a transport cost element. More competitive market conditions in the US provide a further explanation for the relatively low US prices.

However, for the purposes of this chapter, the most important observation is the longer term price trend. From the mid-1980s the prices appeared by and large to be stable, but they were priced in nominal dollars. When adjusted for inflation, using the World Bank Index of International Inflation, the 20-year price history can be summarized as in Table 4.4.

Table 4.3
Prices for Natural Gas in Three Major Markets ($/mmBTU)

	US wellh/HH*	Europe CIF	Japan LNG CIF
1975	0.4	1	1.7
1976	0.6	1.4	1.8
1977	0.8	1.7	2.1
1978	0.9	1.9	2.4
1979	1.2	2.1	2.8
1980	1.6	3	5.1
1981	2	3.6	5.8
1982	2.5	4	5.7
1983	2.6	3.8	5.1
1984	2.7	3.7	4.9
1985	2.6	3.83	5.23
1986	2	3.65	4.1
1987	1.7	2.59	3.35
1988	1.7	2.36	3.34
1989	1.7	2.09	3.28
1990	1.64	2.82	3.64
1991	1.49	3.18	3.99
1992	1.77	2.76	3.62
1993	2.12	2.53	3.52
1994	1.92	2.24	3.18
1995	1.69	2.37	3.46
1996	2.76	2.43	3.66
1997	2.53	2.65	3.91
1998	2.08	2.27	3.05
1999	2.27	1.73	3.14
2000			

Note: *Wellhead until 1988; thereafter Henry Hub.
Source: BP Amoco, June 1990 and June 1999

Expressed in constant 1999 US dollars, the prices in Europe and Japan exhibit significant trend declines over the period under review. The relatively small increase in the real US price in the 1990s is probably explained by the change of location at which the price is measured. This evidence is strongly suggestive that an inadequate resource base or rising cost has not restrained supply in the past.

For a fuller analysis of the adequacy or otherwise of prices to assure future supply, the historically observed prices must be combined with a view to future prices. For example, the World Bank forecasts real gas prices in Europe and North America in 2010 to be no higher than the average prices during the 1990s decade. The same is true of the European Commission's

Table 4.4
Gas Prices (Constant 1999) $/mmBTU

	1979	1989	1999
US (Wellh/Henry Hub)	1.9	1.9	2.3
Europe, CIF	3.3	2.3	1.7
Japan LNG, CIF	4.4	3.6	3.1

Sources: BP Amoco 1980, 1990 and 2000; World Bank 1994 and 2000

view of prices in the European market in 1999. Nevertheless, the stable and historically relatively low real prices recorded since the mid-1980s have induced substantial increases in output during the 1990s, many from new projects, despite a complete absence of signals about impending price increases. This suggests that the cost of supply at the margin can be adequately covered by the prevailing prices.

The above price analysis, along with the earlier observation that reserves of gas are large and growing, justifies the assertion that resources and production will not constrain supply in the future either, and that even a very fast increase in demand can be satisfied at prices not much higher than those experienced in the most recent decade.

Outlook on the Role of Gas in the World Energy System

There is general agreement among energy market analysts that consumption of gas will expand at a pace exceeding that of total primary energy and, therefore, the relative share of gas will continue on its historical upward trend. Table 4.5 presents recent forecasts made by the International Energy Agency and the Energy Information Administration of the US Department of Energy, two reputable energy analysts, with 1999 figures reworked from Table 4.1 shown for reference. Both forecasts envisage a large increase in gas usage with rising shares for this fuel in total primary energy consumption. Both point to a particularly speedy development regarding consumption in the European Union, with North America most lethargic among the geographical regions shown. However, the EIA figures depict a far more dynamic growth for gas, with total consumption in 2020 almost one billion TOE above that projected by the IEA.

Table 4.5
Forecasts of World Gas Consumption

Source and Region	MTOE			Percentage Increase		Share in Total Primary Energy (Percentage)		
	1999	2010	2020	1999–2010	2010–2020	1999	2010	2020
IEA 1998								
European Union	327	506	625	54.7	23.5	23.1	26.0	30.6
FSU and Central Europe	538	647	835	20.3	29.1	46.2	45.3	50.2
USA and Canada	620	705	676	13.7	–4.1	25.5	25.9	23.8
Japan, Australia, NZ	89	119	132	33.7	10.9	14.3	15.8	16.2
World Total	**2064**	**2721**	**3468**	**31.8**	**27.5**	**24.2**	**23.6**	**25.2**
EIA 2000								
European Union	327	499	653	52.6	30.9	23.1	27.3	33.0
FSU and Central Europe	538	786	1038	46.1	32.1	46.2	49.5	54.4
USA and Canada	620	794	930	28.1	17.1	25.5	24.9	26.9
Japan, Australia, NZ	89	121	141	36.0	16.5	14.3	15.4	16.9
World Total	**2064**	**3218**	**4367**	**55.9**	**35.7**	**24.2**	**25.5**	**28.5**

Sources: BP Amoco 2000; EIA 2000; IEA 1998a

Among the geographical areas listed, the European Union and Japan will have to rely on increasing imports to provide for the envisaged expansion in gas consumption. This will require substantial investments in transport infrastructure, pipelines in Europe and LNG facilities on the supply routes to Japan. Such investments are well under way, as frequently reported in the commercial press. New Russian pipelines have been laid across Poland and a northern route is being planned via an extension of the existing network to Finland. Additional LNG capacity to provide for the needs of Japan and other Asian countries is under construction in northern Australia, to supplement the traditional suppliers in Indonesia and Malaysia. To assure a satisfactory degree of supply security, the importers make efforts to diversify their supplies. The Europeans provide political support to the expansion of deliveries from Algeria and Central Asia, while Japan is keen to increase the role of the Middle East as a provider of its gas needs.

Why has gas increased its share in world energy markets over time? What are the factors behind its anticipated continued relative expansion in future? Are the forecasts contained in Table 4.5 credible? Are they overly euphoric or conservative about the future of gas? The next section tries to provide some answers to these questions.

The Increasing Competitiveness of Gas

A number of factors have combined to improve the competitiveness of gas as an energy fuel over the past two decades. The influence of these factors is far from spent and will continue to favor gas in the coming years. A majority of these factors have to do with the use of gas to generate power, the most important growth area for gas.

Impressive Reductions in the Cost of Supply

Reliable figures on the cost of production and transport are hard to find. Gas producers and their consultants, who may know, have a self-interest in showing that supply carries a high cost, for that helps them to negotiate better prices. Despite the uncertainties, two aspects strongly suggest a reduced supply cost over time.

The first aspect relates to the most recent additions to European supplies from Nigeria as LNG, and from the Russian Yamal Peninsula through a new pipeline system via Poland, both very distant sources. In the early 1990s the profitability of these sources was commonly believed to require substantially higher prices than those realized in the past decade or forecast for the present one. Apparently, this is no longer so.

The profit calculations underlying the investments in the LNG facilities in Nigeria, with production starting in 1999 and with distant Europe as the primary market, must have been based on a delivery price no higher than $2.50/mmBTU, the average nominal dollar price in Europe during the decade, with no long-term increase in sight. The same must be true of the new Russian supplies, originating at a long distance from the ultimate markets in Western Europe. Therefore, the cost of delivery in both cases must have declined to below $2.50/mmBTU or otherwise there would have been little incentive to invest in the ventures.

The second aspect is the analysis of the International Energy Agency regarding cost-reducing technological progress and managerial improvement. A general conclusion from a detailed analysis of the cost of gas supply is that "recent development seems to indicate that there is still a huge untapped potential for lowering production costs." Later on, the study discusses the North Sea, but its findings have a strong relevance for offshore production anywhere: "Recent studies for production in the North Sea indicate that production costs could be lowered by 30 to 50 percent in the future. This could be achieved both by a wider use of new technologies, as well as new management approaches."[2] A more recent IEA study deals with the cost of oil production. It notes that the combination of competitive cost pressures and technological progress have reduced the world average finding and capital costs from $12/barrel in 1981 to $4 in 1996, while the operating costs were cut by half in the same period.[3] The findings undoubtedly have a clear bearing on the cost of gas production too.

The Impact of the Oil Price Collapse of the mid-1980s

In most markets (though less so in the United States) the price of gas has traditionally been linked to competing energy sources, predominantly oil products. Therefore, gas prices declined sharply in 1986 as a result of the oil price collapse in that year. While the competitiveness of gas prices remained unchanged vis-à-vis oil, gas prices rose in those energy markets where gas competed with other fuels, e.g. coal, nuclear and biomass, whose supply prices do not ordinarily follow the movements of the oil price. As a result, gas became more competitive than coal or nuclear fuels regarding the generation of power. Regarding the generation of heat, gas became more competitive than coal and biomass. Gas also became more competitive than electricity, whose supply continues to be dominated by sources other than oil and gas.

Deregulation and Reduced Monopoly Power in Gas Supply

In both Europe and Japan the supply of gas contains significant monopolistic elements. By linking the prices for their deliveries to the price of oil, gas suppliers reaped very substantial monopolistic rents while the oil prices remained high. After the collapse of oil prices these rents shrank, but some monopoly gains nevertheless remained. Deregulation in the UK in the early 1990s introduced competition in the supply of gas, which changed the configuration of the market fundamentally. Prices fell promptly to levels significantly below those in continental Europe.

Competition in the energy sector has come much later to the European continent. Market forces loosened up the monopoly power of the suppliers. Exporters to Europe became frustrated with the slow market growth in the 1980s and some competition between them emerged in the course of the 1990s. Wingas, a joint venture between BASF, a German chemical giant with substantial gas consumption, and Gazprom, the Russian gas monopoly, aggressively built a pipeline system parallel to that of Ruhrgas, the dominant German gas player. This feat introduced some competition in Europe's most important market. Such competition was accentuated toward the end of the decade, when the Interconnector, a gas pipe across the English Channel, provided an outlet for deregulated and cheap British gas to the continental market. Competition in Europe is bound to make further progress with the introduction of the European gas directive in mid-2000, assuring third-party access to the existing pipeline network. As a result, most of the remaining monopolistic features are likely to crumble, along with the traditional price link between oil and gas. Gas should become somewhat cheaper than it would otherwise have been as these developments mature, and its ability to conquer a larger share in the energy market will improve.

Talk about deregulation of the energy market is going on in Japan too, but so far action has been limited. Competition in the power sector, now characterized by regional monopolies, will certainly put pressure on the gas suppliers to lower prices, since such competition will make it much harder to pass on a high price to consumers. With reduced import prices for gas, other markets besides that for power might open up too, thus promoting an expanded gas share.

Environmental Advantage

The environmental disadvantage of fossil fuels stems in particular from the air emissions that arise from their use. Traditionally, SO_2, NO_x and soot have been the focus of concern, but more recently greenhouse gas emissions have dominated the environmental agenda. The benign feature of gas is that its burning generates very little SO_2, NO_x and soot, if any at all. In that respect, it compares very favorably with oil, coal and even biomass.

However, gas is also favored because of its limited carbon content. An index of CO_2 emissions per unit of energy, setting coal at 100, yields roughly 71 for oil and 57 for gas, although the figures may vary depending on the features and uses of the gas. A common key feature in proposed implementations of the Kyoto agreement is to substitute gas for coal on a large scale, thereby significantly reducing carbon emissions for a given volume of energy consumed. If this situation became reality, then the share of gas in world energy would undoubtedly rise beyond the IEA and EIA projections of Table 4.5.

A less discussed aspect of gas in this context is the leakage of methane from inadequately maintained pipes, a common feature in the FSU. Methane's climate potency is many times greater than that of CO_2, even with its shorter life time taken into account, so the climate advantage of gas could easily be nullified or even reversed if such leakage is accounted for. The repair of the pipes ought not to impose an overwhelming burden on the gas industry and, by saving the fuel, could be economical in its own right.

The sensitivity to the environment in the industrialized world has clearly improved the competitive chances of gas in past decades. This sensitivity continues to grow worldwide. Gas is most likely to benefit in the process.

Repeal of Directives against Gas Use in Power Generation

During the 1970s the US administration and the European Commission implemented directives against the use of gas in the generation of power. The rationale for these directives was a perception that gas is a scarce and superior resource, to be preserved for high value uses, for example by households or as raw material in the chemical industry. It is not clear how effective the directives have been. In the Netherlands, for instance, the share of gas in electricity generation increased throughout the period that the European directive was in force. Nevertheless, at least in Europe, the directive played into the hands of the gas suppliers, for it facilitated their monopolistic goals of charging high prices for limited supplies. The directives were repealed both in the US and Europe in the late 1980s, as it became increasingly clear that gas was a truly abundant resource, the wider use of which could significantly reduce the cost of electricity. This opened up a huge market for gas, whose potential had only begun to be understood.

Privatization of the Energy Sector and the Cost of Capital

The world's energy sectors are being swept by a wave of privatization or at least commercialization. This has interesting positive repercussions for gas. A characteristic of gas is that its use requires fewer capital-intensive structures to produce power than the use of alternative fuels. When gas was competing with other fuels in the energy market, low capital costs traditionally compensated for relatively high fuel costs. Until recent times, however, in many parts of the world capital was frequently provided to the energy sector by the government at below market rates. As a result, the advantage of low capital-intensity gas was suppressed. However, things are changing. The energy sector's access to cheap finance is drying up, as governments abdicate their central roles in energy supply throughout the world. The increasing cost of finance, procured at commercial rates, is a further factor that strengthens the competitiveness of gas.

Therefore, while the ongoing commercialization in the energy sector is likely to favor gas consumption for the reasons mentioned, the new market situation of gas consumers may involve some discouragement to the suppliers of gas. Until recently, most power generators were shielded from competition and could pass on any cost increases to the end-users of electricity. They were thus willing to enter into long-term supply contracts with gas producers at predetermined volumes and prices. Since investments in gas production and transport often require very large amounts of capital, to be written off over extended periods, such contractual arrangements favored the development of a gas supply.

With the transformation of the energy markets, the producers of power are increasingly exposed to the vagaries of competition. While they may guarantee volumes, they are unwilling and unable to ensure price levels over time. Therefore, gas producers and their financiers have to carry more of the price risk than before. However, this factor should not be exaggerated. Virtually all other raw material suppliers are also exposed to the risk of fluctuating prices.

Technological Improvements in Gas Usage

Technological progress that reduces the cost of gas supply is matched by cost-reducing improvements in the use of gas supply. The most spectacular development in the past 20 years has been the commercial breakthrough of the Combined Cycle Gas Turbine (CCGT) for power generation. The CCGT achieves efficiency rates of more than 60 percent, compared to the rates of 40 percent that were common in conventional gas steam turbines and to the slow improvement of rates to 35–40 percent in coal plants. This breakthrough has made gas superior to coal over a much wider range of relative prices. When new power plants are to be built, power production fueled by gas has become the economically self-evident choice in almost all circumstances where gas supply is available.

Small Minimum Economic Size Power Generation Plant

Economies of scale are important for power plants using coal and nuclear fuels, but not for plants that burn gas. The minimum economic size of a power plant using gas is only a fraction of the required size when the other fuels are used. This difference was unimportant when demand for energy was growing at fast rates, for the output from even very large new units was easily and quickly absorbed by an expanding electricity market. Since the 1990s, however, demand for energy has been growing at very slow rates in the mature industrialized economies. Phasing in large new units has become more cumbersome and less economic under the stagnant market conditions. The possibility of producing electricity in small gas-fueled units has afforded gas yet another competitive advantage of increasing importance during the 1990s. This factor has less significance in many industrializing economies, where the demand for energy is fast expanding.

Deterrents to Gas Expansion

Two factors restrain the pace and extent of gas expansion. The first one is the longevity of existing installations. The viability of an energy plant in operation is based on its capital cost. New plants will be able to compete only when their total cost, including the cost of investments, is lower than the operating and fuel costs of plants already in operation. This is only seldom the case. Operators therefore have a strong incentive to extend the life of existing units almost in perpetuity. Experience in the US power industry reveals the difficulties gas faces in making strong inroads into a sector dominated by coal, despite a clear superiority in terms of total costs. When properly maintained, coal and nuclear plants can easily continue economic operations for 50 or 60 years and even longer. Hence, when energy grows at slow rates, so that there is little need for new, as distinct from replacement, capacity, the scope for structural change will be limited.

The second factor concerns supply security, an issue that was in the vanguard of energy policy in major consumer countries until the mid-1980s but that has been a bit out of fashion since then. The factor has more relevance for the European nations, Japan and Korea, whose gas needs are supplied from abroad in great measure. With the experiences from the 1970s of the havoc caused by the oil crises, these nations' politicians may become wary of increasing the share of gas far above recent levels, given the concentration of supply with a few exporters and the political uncertainties that characterize some of these suppliers, e.g. Algeria, Russia and Indonesia. In addition to the attempts to diversify the sources of supply, discussed in an earlier section, it is conceivable that energy policies might emerge in importing countries to restrain an excessive expansion of gas.

Concluding Observations

There is no way in which the impact of the factors briefly discussed above could be quantified. Only qualitative judgments are possible. Despite the reservations expressed, even the faster growth of volume and share projected by the EIA (Table 4.5) is likely to be an underestimate of the future role of gas. The statement in the introduction can therefore be repeated: the future of natural gas in the world energy market is quite bright.

The Maverick: From Gas to Middle Distillates

It is also necessary to speculate about a possible maverick that might provide a further boost to gas. Gas is delivered to consumers either by pipe or as LNG. Piped gas is exceedingly bulky, while LNG requires elaborate installations for liquefaction and regasification and expensive cryogenic ships for the ocean voyage. The cost of transporting this fuel through either shipping mode is high and typically accounts for a large proportion of total delivered costs in the world's major markets. The location of the gas deposits then becomes an important factor for their wellhead value. In numerous offside locations this value is zero and the deposits are never developed, for the sum of the cost of production and transport exceeds the price at which gas competes with other fuels in final markets.

The transformation of gas to fuels that are liquid at room temperature would greatly reduce the cost of shipping and potentially open up gas regions that would remain uneconomical with the traditional means of transport. Technically, the problem of converting gas to liquids, which for all practical purposes are identical to oil middle distillates like kerosene or gasoil, has been solved, but the economics of such conversion have been marginal at best.

The first commercial-size gas-to-middle-distillate plant was established by Shell in Bintulu, Malaysia, and started operations in the early 1990s. Its capacity was 12000–15000 barrels of middle distillates per day. The products from the plant commanded a premium price compared with their oil-based substitutes. On account of their purity, the gas-based products could be blended with oil-based distillates to improve their quality or sold to markets with particular environmental sensitivity. Even so, they did not provide sufficient financial return, so the company turned part of the output into specialty products such as waxes, solvents and detergent feedstocks, with higher added value to improve profitability. Some of the markets for these specialty products are quite small and became virtually saturated by the output from Bintulu.

In the late 1990s Exxon, Shell, Statoil and others revealed investment plans for plants on a larger scale based on more modern technology, and with a seemingly promising profit potential, to use the rich and underutilized gas resources in Qatar and off northern Australia, amongst others.

Nevertheless, the technology still has not been commercially successful on a large scale. Important factors in the economics of gas-to-middle-distillate conversion comprise the price at which the gas can be procured, the efficiency of conversion, the cost of conversion and the price of crude petroleum and products. In contrast, the cost of transporting the output is a secondary consideration, as is the case for oil and petroleum products.

Clearly, gas-based middle-distillate plants will be uneconomical where the gas has to be bought at a high price because it can be sold directly into nearby energy markets. This is not so in Qatar, where gas is exceedingly abundant but its use is limited by the small demand for piped supplies, domestically and in nearby countries, and as a feedstock for LNG. The situation is somewhat similar for the offshore deposits in northern Australia. The identified gas wealth has grown in strides, but the demand is limited by the capacity of the LNG plants, which is in turn constrained by oversupplied Far Eastern markets. In both places, therefore, gas can be procured at low cost. The cost could be even lower for undeveloped gas deposits without any alternative market outlet, either because they are awkwardly located or because they are too small to motivate an investment in the means of transport. The economic viability of gas-based middle-distillate plants requires a very careful choice of sources of gas that can be procured at low cost.

The efficiency and cost of conversion are clearly improving as a result of ongoing technological development. Traditionally, a lot of energy was lost during conversion. Exxon claims for its planned Qatari plant[4] that 60 percent of the energy input will be extracted, twice the extraction of Shell's older Bintulu installation, but still far below the 85 percent efficiency of conversion of LNG. Technical improvements and economies of scale also promise a lower unit cost than in the older plant.

The relatively low average prices of oil, namely $15–20 per barrel during the 1990s, have imposed an effective cap on the profitability of gas-to-middle-distillate operations, for the output has to compete directly with oil products. At oil prices above $25, the gas-based middle distillates would surely be a great financial success, but so would many other forms of energy that cannot compete with cheap oil. If reputable price forecasters like the IEA or the World Bank are right, then the high prices of 2000 constitute a temporary spike, and there will be little support from this quarter for the gas-to-middle-distillate industry in the longer term.

Therefore, a commercial breakthrough will depend on further technical improvements that increase the efficiency rates and lower the transformation costs. It is unclear whether and when such a breakthrough will occur.

Once it does, it could become a true maverick with the potential for transforming both the oil and the natural gas markets. The oil market would be transformed by an added source of supply with high potential and a very broad geographical spread. As a result, the fear of oil depletion should subside. Upward pressures on oil prices, from depletion concerns or from market action by producers, will be weakened as a result. Given the purity of the gas-based middle distillates, the average environmental impact of oil products will be improved.

The gas market will be transformed in more direct and fundamental ways. First, the economic resource base will be greatly expanded, making it worthwhile to exploit many deposits that are not competitive under currently prevailing conditions. This, in turn, will prompt more exploration in areas where no one has bothered to look for gas. As a result, the quantity of the resource is bound to expand. Second, the market for gas will be redefined and greatly expanded. Traditionally, demand for gas has been constrained by this product's characteristics, both in terms of uses and geography. Gas has conquered important market segments in the power and heat sectors, and supplies energy to households and industrial firms. However, it has hitherto had little economic justification in the important transport sector. Gas usage has been geographically limited to densely populated areas where consumption is sufficient to motivate the high investment in the transport and distribution of gas. Both constraints will disappear once the transformation of gas into middle distillates becomes commercial. Gas will then compete directly in virtually all markets that had earlier been exclusive oil preserves. The easy transport of middle distillates will break the geographical barriers for gas use too.

In this way, gas will have the potential to expand its share in world energy consumption far beyond even the most optimistic projections, such as those presented in Table 4.5 above. At the same time, the distinction between the gas and the oil market will become increasingly blurred. At the final consumer end it will be hard to determine whether gas or oil is being used.

5

The Importance of Energy Quality in the Transition to a Solar Economy

Cutler J. Cleveland

G lobal oil production will peak in the coming decades, followed by natural gas and coal. These turning points constitute an unprecedented watershed in human history. This chapter focuses on some of the critical challenges we face in the transition from conventional fossil fuels to alternative sources, particularly solar energy. Foremost is the capacity for renewable fuels to develop into the functional equivalent of fossil fuels, i.e. to have a similar capacity to generate goods and services per unit of energy input. The cost of many renewable energy systems has declined in the past two decades, but significant challenges remain. This chapter describes the nature of some of those challenges, especially with regard to the prospects for overcoming the spatially diffuse nature of many renewable energy systems. It also explores the relationship between energy use and economic growth, and the factors that determine the strength of that relationship.

Energy Transitions in the Past

The level of health, food security and especially material standard of living that exists today throughout the world is made possible by the extensive use of fossil fuels. While many take this affluence for granted, a long-term view illustrates that the fossil fuel era is relatively new and will last for a relatively short period of time (see Figure 5.1). For thousands of years prior to the Industrial Revolution, human societies got their energy from the products of photosynthesis, principally fuel wood and charcoal. Widespread use of coal did not develop until the eighteenth century, oil and gas not until the late nineteenth century.

The history of energy use in the United States illustrates these transitions (Figure 5.2). In 1800 the source of energy was animal food, which provided energy to the draft animals on farms, and wood fuel, which was used for domestic heating and cooking and by early industry. The Industrial Revolution transformed the nation's energy picture, substituting coal for renewable fuels on a massive scale. By the time of the First World War, coal accounted for nearly three quarters of the nation's energy use. Wood and animal feed were rapidly disappearing, the latter due to the introduction of the first tractor in 1911.

However, coal's position as the dominant fuel was fleeting. Oil and natural gas quickly replaced coal, just as coal had replaced wood. By the 1960s

Figure 5.1
The Epoch of Fossil Fuels in Human History

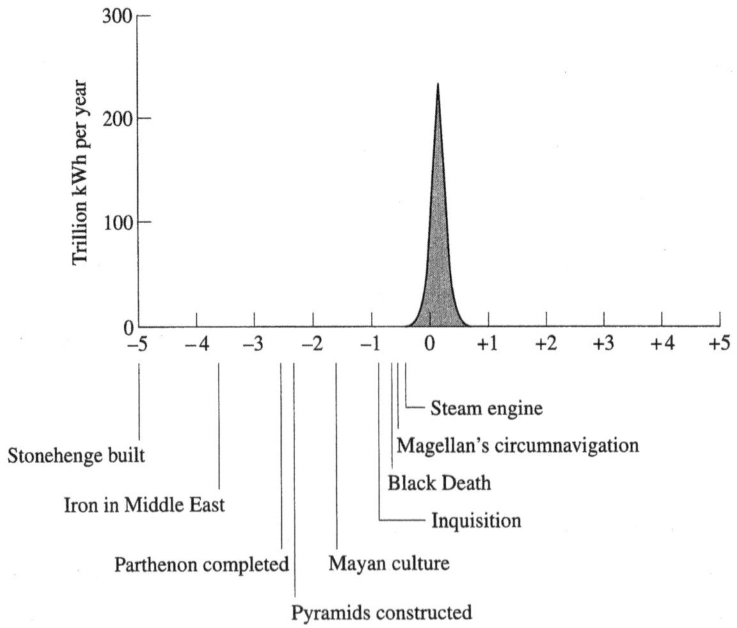

Source: Hubbert 1969

[130]

Figure 5.2
The Composition of US Energy Use

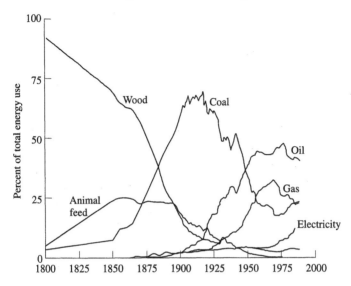

Source: Hall et al. 1986

oil and gas together accounted for more than 70 percent of total energy use; coal had dropped to less than 20 percent. Primary electricity has played a small but steadily growing role. Primary electricity refers to electricity generated by hydroelectric, nuclear, geothermal, solar and other so-called "primary" sources. The increase in the share of primary electricity toward the end of the period is due to the rise in nuclear generating capacity.

Energy Return on Investment

This long-term view of energy raises an important question: what guided these transitions in the past and to what extent can such information inform us about the impending transition from fossil to renewable fuels? The transition from one major energy system to the next is driven by a combination of energy, economic, technological and institutional factors. The energy-related forces stem from the tremendous economic and social opportunities offered by new fuels and their associated energy converters, especially in comparison to earlier fuels.

A key aspect of an energy delivery system is its energy return on investment (EROI).[1] The EROI is the ratio of the gross energy extracted to the energy used in the extraction process itself (Figure 5.3). A related term is net energy or energy surplus, which is equal to the gross energy extracted minus the energy used in the extraction process. Thus, the EROI is a ratio reflecting the return on energy investment, while energy surplus is a quantity of energy delivered to the rest of a system after the energy cost of obtaining it has been paid.

Concepts were developed in ecology to describe the critical role energy plays in nature. All organisms must use energy to perform a number of life-sustaining tasks such as growth, reproduction and defense from predators. The most fundamental task of all is using energy to obtain more energy from the environment. When energy is used to do useful work, energy is degraded from a useful, high-quality state to a less useful low quality state. This means that all systems must continuously replace the energy they use, and to do so takes energy. This fundamental reality means that EROI and net energy are used to explain the foraging behavior of organisms, the distribution and abundance of organisms[2] and the structure and functioning of ecosystems.[3]

While human society is driven by much more than simple energy imperatives, the concepts of net energy and EROI help to explain the dramatic energy transitions of the past. For the overwhelming period of their existence, humans obtained energy from the environment by hunting and gathering. The EROI for food capture is the caloric value of the food capture to the expenditure of energy in the capture or gathering process. The EROI for energy-dense roots is 30 to 40; a reasonable average for all gathering is 10

Figure 5.3
The Energy Return on Investment (EROI)

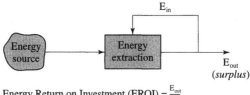

Energy Return on Investment (EROI) $= \dfrac{E_{out}}{E_{in}}$

Source: Hall et al. 1986

to 20 (Table 5.1).[4] The shift to agriculture represented a fundamental shift in the way humans obtained food energy from the environment. Agriculture required greater inputs of energy compared to hunting and gathering. The forest had to be cut or the wetland had to be drained to free the land for cultivation. The land had to be prepared for planting, the crop had to be planted, cared for and ultimately harvested. All of these activities required substantial inputs of energy. As a result, the EROI for agriculture often was less than or about equal to that for hunting and gathering (Table 5.1).

Table 5.1
EROI for Food Production

System	EROI
Hunter-Gatherer	
• Energy-dense roots	30–40
• All gathering	10–20
• Hunting – large ungulates	10–20
• Fishing – coastal whaling	2,000
Agriculture	
• Shifting cultivation	10–40
• Traditional agriculture	10–30
• Industrial – crops	<10
• Industrial – animals	<1

Source: Smil 1991

From an energy perspective, why did agriculture replace hunting and gathering? The answer lies in the size of the energy surplus delivered by agriculture. Although the EROI for agriculture may have been lower, the energy inputs increased faster than the EROI declined, to such an extent that surplus food energy increased. The greater surplus vastly increased the capacity of the environment to provide for human beings and offered new economic, social and cultural opportunities. Natural ecosystems produce enough edible food energy to support hunter-gatherers at densities no greater than one person per square kilometer. Traditional agricultural societies support hundreds of people per square kilometer, enabling permanent settlements to grow in size and number. The greater surplus released labor from the land, creating the potential for people to move to urban areas and work in manufacturing and industry.

Energy Converters

Simultaneously with a change in the types of energy used, there was a change in the machines used to convert that energy into useful work. The economic usefulness of an energy converter is determined in part by its power, the rate at which it converts energy to do useful work. In economic terms useful work refers to the use of energy to produce goods and services.

Animate energy converters (humans and draft animals) convert energy while producing low energy outputs. The energetic limits of people and draft animals set very definite economic and social bounds. The Industrial Revolution erased these limits with the introduction of the steam engine, which had a power output that dwarfed that of animate sources. The higher power output of the steam engine enabled it to deliver a much larger energy surplus than human labor or draft animals. The higher energy surplus expanded economic opportunities just as agriculture had done compared to hunting and gathering, only on a much grander scale.

Given the economic advantage offered by heat engines powered by fossil fuels, it is no surprise that heat engines rapidly replaced labor and draft animals once they became available. The United States' economy illustrates this transition. In 1850 human labor and draft animals did more than 90 percent of the work in the economy. Over the next half century, engines powered by wood and then coal rapidly displaced the animate converters. By the 1950s labor and animals had been almost completely displaced.

Of those economic changes driven by the new fuels and machines, one of the most dramatic was the effect on labor productivity. In agriculture, for example, the productivity of labor increased more than a hundred-fold relative to rates possible prior to the Industrial Revolution. As mentioned above, this increase in labor productivity reduced the need for labor in the agricultural sector, thus providing a potential supply for burgeoning industrial sectors. The same increase in labor productivity occurred in other sectors of the economy where the efforts of human labor were subsidized by more energy and more powerful energy converters. Indeed, labor productivity in general is related to the amount of energy used per laborer.[5]

Energy and Economic Growth

How strong is the connection between energy use and economic growth? There are quite divergent views on this subject. One hypothesis is that the link is weak, that the production of goods and services can be decoupled from energy inputs. There are several forces that might achieve this. First, it generally is assumed that as fossil fuels become scarcer, their price will rise, which in turn will trigger technological changes and substitutions that improve energy efficiency. Indeed, many believe that the price shocks in 1973–74 and 1979–80 led to the adoption of many new energy-efficient technologies. Second, the shift to a service-oriented, dot-com economy will decouple energy use from economic activity. A dollar's worth of steel requires 93,000 BTU to produce in the United States; a dollar's worth of financial services uses 9,500 BTU. Thus, it stands to reason that a shift toward less energy-intensive activities will reduce the need for energy. Third, growing environmental imperatives will create new incentives to decouple energy use from economic output. In particular, concern over climate change is increasing the need to decarbonize the economy. Improving the efficiency of energy use is one way to do this.

International comparisons of energy use seem to lend support to the hypothesis that energy use and economic growth are weakly linked.

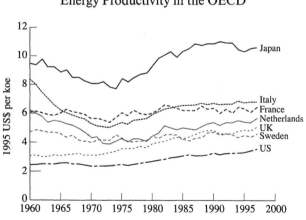

Figure 5.4
Energy Productivity in the OECD

Source: World Bank Indicators 1997

Aggregate energy efficiency is frequently measured by the energy/real GDP (E/GDP) ratio. This is the quantity of commercial energy use compared to the quantity of GDP produced. By this measure, many industrial nations have become more energy efficient in the last few decades (Figure 5.4). The E/GDP ratio declined even faster after the energy price shocks, lending support to the argument that the price increases triggered energy-saving technological change. Note also that the level of energy efficiency varies significantly among nations. This suggests that nations such as the United States have the potential to improve their energy efficiency while maintaining the same level of output.

Energy Quality and the Energy/GDP Ratio

A second hypothesis is that the connection between energy use and economic output is strong. A key concept in understanding this argument is energy quality. From an economic perspective, the value of a heat equivalent of fuel is determined by its price. The value marginal product of a fuel in production is the marginal increase in the quantity of a good or service produced by the use of one additional heat unit of fuel multiplied by the price of that good or service.[6]

The marginal product of a fuel is determined in part by a complex set of attributes unique to each fuel such as physical scarcity, capacity to do useful work, energy density, cleanliness, amenability to storage, safety, flexibility of use, cost of conversion and so on. However, the marginal product is not uniquely fixed by these attributes. Rather, the energy sector's marginal product varies according to the activities in which it is used, how much and what form of capital, labor and materials it is used in conjunction with, and how much energy is used in each application. As the price rises due to changes on the supply side, users can reduce their use of that form of energy in each activity, increase the amount and sophistication of capital or labor used in conjunction with the fuel, or stop using that form of energy for lower value activities. All these actions raise the marginal productivity of the fuel. When capital stocks have to be adjusted, this response may be somewhat sluggish and lead to lags between price changes and changes in the value of the marginal product.

The heat equivalent of a fuel is just one of the attributes of the fuel and ignores the context in which the fuel is used, and thus cannot explain, for example, why a thermal equivalent of oil is more useful in many tasks than a heat equivalent of coal.[7] In addition to attributes of the fuel, a marginal product also depends on the state of technology, the level of other inputs and other factors. According to neoclassical theory, the price per heat equivalent of fuel should equal its value marginal product and, therefore, represent its economic usefulness. In theory, the market price of a fuel reflects the myriad factors that determine the economic usefulness of a fuel from the perspective of the end-user.

Consistent with this perspective, the price per heat equivalent of fuel varies substantially among fuel types (Table 5.2). The different prices demonstrate that end-users are concerned with attributes other than heat content. As Berndt states:

> Because of [the] variation in attributes among energy types, the various fuels and electricity are less than perfectly substitutable – either in production or consumption. For example, from the point of view of the end-user, a Btu of coal is not perfectly substitutable with a Btu of electricity; since the electricity is cleaner, lighter, and of higher quality, most end-users are willing to pay a premium price per Btu of electricity. However, coal and electricity are substitutable to a limited extent, since if the premium price for electricity were too high, a substantial number of industrial users might switch to coal. Alternatively, if only heat content mattered and if all energy types were then perfectly substitutable, the market would tend to price all energy types at the same price per Btu.[8]

Do market signals (i.e. prices) accurately reflect the marginal product of inputs? Kaufmann (1994) investigates this question in an empirical analysis of the relation between relative marginal product and price in US energy markets.[9] To do so, he estimates a reduced form of a production function that represents how the fraction of total energy use from coal, oil, natural gas and primary electricity (electricity from hydro and nuclear sources) affects the quantity of energy required to produce a given level of output. The partial derivatives of the production function with respect to each of the fuels gives the marginal product of individual fuels, in which marginal product is defined as the change in economic output given a change in the use of a heat unit of an individual fuel. The equations are used to calculate the marginal product for each fuel type for each year between 1955 and 1992.

[137]

Table 5.2
US Market Price for Various Energy Types

Energy Type	Market Price ($/10^6 BTU)
Coal	
Bituminous	
• mine-mouth	0.91
• consumer cost	1.40
Anthracite	
• mine-mouth	1.43
Oil	
• wellhead	2.27
• distillate oil	7.03
• jet fuel	3.95
• LPG	6.22
• motor gasoline	8.91
• residual fuel oil	2.32
Biofuels	
• consumer cost	1.88
Natural Gas	
• wellhead	1.65
• consumer cost	4.15
Electricity	
• consumer cost	20.34

Source: Cleveland et al. 2000. Values are 1994 prices

Figure 5.5
GDP and Energy Use in the USA

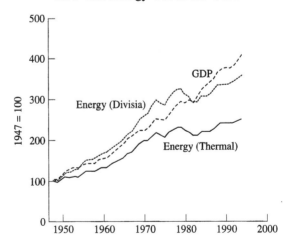

Source: Cleveland et al. 2000

[138]

The time series for marginal products are compared among fuels and these ratios are related to relative prices using a partial adjustment model. The results indicate that there is a long-term relation between relative marginal product and relative price, and that several years of adjustment are needed to bring this relation into equilibrium. This suggests that over time prices do reflect the marginal product – and hence the economic usefulness – of fuels.

Other analysts calculate the average product of fuels, which is a close proxy for marginal products. Adams and Miovic estimate a pooled annual cross-section regression model of industrial output as a function of fuel use in seven European economies from 1950 to 1962.[10] Their results indicate that petroleum is 1.6 to 2.7 times more productive than coal in producing industrial output. Electricity is 2.7 to 14.3 times more productive than coal. Using a regression model of the energy/GDP ratio in the US, Cleveland found that the quality factors of petroleum and electricity relative to coal were 1.9 and 18.3, respectively.[11]

Energy quality is important to account for in the assessment of E/GDP ratios.[12] When energy use is calculated in the standard heat equivalents, energy use and GDP diverge in the United States, seemingly consistent with the de-linking hypothesis. Figure 5.5 also shows energy use represented in a Divisia index, a method for aggregating heat equivalents by their relative prices. This quality-corrected measure of energy use shows a much stronger connection with GDP. This visual observation is corroborated by econometric analysis that confirms a strong connection between energy use and GDP when energy quality is accounted for.[13]

International comparisons of the E/GDP relationship also demonstrate the importance of energy quality. Econometric analysis of the E/GDP in the United States, Japan, the UK and France since 1950 indicates that changes in energy quality explain much of the variation.[14] Declines in the E/GDP ratio are associated with the general shift from coal to oil, gas and primary electricity. Also important are fuel prices, the structure of economies and household purchases of energy.

The Quality of Solar Energy

There is no shortage of energy on Earth (Table 5.3). Indeed, the storages and flows of energy on the planet are staggering in relation to human needs.

Table 5.3
US Energy Stocks and Flows

Source	Stock (10^{18} BTU)	Flow (10^{15} BTU per yr)	Multiple of Current US Annual Energy Use
Coal	80		849
Oil	17		179
Oil Shale	22		228
Tar Sands	0.2		2
Gas	7		72
Unconventional Gas	5		54
Gas Hydrates	206		2,187
Uranium	3		29
Incident Solar Energy		46,700	496
Geothermal		39,685	421
Wind		155	2
Hydrogen		78	0.83
Biomass		47	0.50
Hydropower		2	0.02
Tides		1	0.01

Source: Hall et al. 1986

Consider the following facts:

- The amount of solar energy intercepted by the Earth every *minute* is greater than the amount of fossil fuel the world uses every *year*.
- Tropical oceans absorb 5.3×10^{20} BTU of solar energy each year, equivalent to 1,600 times the world's annual energy use.
- The potential energy in the winds that blow across the United States each year could produce more than 4.4 trillion kWh of electricity – more than one-and-a-half times the electricity consumed in the United States in 2000.
- Annual photosynthesis by vegetation in the United States is 4.7×10^{16} BTU, equivalent to nearly 60 percent of the nation's annual fossil-fuel use.

In contrast to its vast quantity, the quality of solar energy is low when compared to that of fossil fuels. Consider the energy flow in the Earth's crust. The total heat loss from the Earth's crust is 1.3×10^{18} BTU per year, equivalent to nearly four times the world's annual energy use. However, this energy flow is spread over the entire 5.1×10^{14} square meters of the Earth's surface. This means that the amount of energy flow per unit area is 2,400 BTU per square meter, an amount equivalent to just 1/100 of a gallon of gasoline.

Consider incoming solar energy. The land area of the lower 48 states of the United States intercepts 4.7×10^{19} BTU per year, equivalent to 500 times the nation's annual energy use. However, that energy is spread over nearly 3 million square miles of land area, so that the energy absorbed per unit area is just 1.5×10^{13} BTU per square mile per year. Plants, on average, capture only about 0.1 percent of the solar energy reaching the Earth. This means that the actual plant biomass production in the United States is just 1.6×10^{10} BTU per square mile per year.

These examples illustrate that heat flow from the Earth, solar energy, plant biomass and other renewable forms of energy are diffuse forms of energy, particularly when we compare them to fossil fuels. This is captured by the concept of power density. Power density combines two attributes of energy sources: the rate at which energy can be produced from the source and the geographic area covered by the source. A coal mine in China, for example, can produce upwards of 10,000 watts per square meter of the mine. As the above examples indicate, most solar technologies have low power densities compared to fossil fuels.

A low energy or power density means that large amounts of capital, labor, energy and materials must be used to collect, concentrate and deliver solar energy to users. This situation tends to make solar energy more expensive than fossil fuels. The difference between solar and fossil energy is best represented by their respective energy return on investment (EROI). The EROI for fossil fuels tends to be large while that for solar power tends to be low (Figure 5.6). This is the principal reason why humans aggressively developed fossil fuels in the first place.

Fossil fuels have allowed us to develop lifestyles that also are very energy intensive. The places that we live, work and shop in all have high power densities. Supermarkets, office buildings and private residences in industrial nations demand huge amounts of energy. This very energy-intensive way of living, working and playing has been made possible by fossil-fuel sources that are just as concentrated.

Another quality difference between renewable fuels and fossil fuels is their energy density, i.e. the quantity of energy contained per unit mass of a

Figure 5.6
EROI for Energy Systems

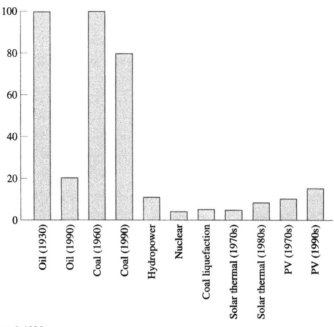

Source: Hall et al. 1986

Table 5.4
Energy Densities of Fuels

Fuel	Energy Density (Mj/kg)
Peats, green wood, grasses	5.0–10.0
Crop residues, air-dried wood	12.0–15.0
Bituminous coals	18.0–25.0
Charcoal, anthracite coals	18.0–32.0
Crude oils	40.0–44.0

Source: Smil 1991

fuel (Table 5.4). For example, wood contains 15 Mj per kilogram; oil contains up to 44 Mj per kilogram. Higher energy densities also contribute to the higher EROI for fossil fuels relative to many renewable fuels.

Conclusions

Among the countless technologies humans have developed, only two have increased our power over the environment in an essential way. Georgescu-Roegen called these Promethean technologies. Prometheus I was fire, unique because it is a qualitative conversion of energy (chemical to thermal) and because it generates a chain reaction that continues for as long as sufficient fuel is forthcoming. As Georgescu-Roegen described its significance:

> The mastery of fire enabled man not only to keep warm and cook the food, but above all to smelt and forge metals, and to bake bricks, ceramics, and lime. No wonder that the ancient Greeks attributed to Prometheus (a demigod, not a mortal) the bringing of fire to us.[15]

Prometheus II was the heat engine. Like fire, heat engines achieve a qualitative conversion of energy (heat into mechanical work), and they sustain a chain reaction process by supplying surplus energy. Surplus energy (or net energy) is the gross energy extracted less the energy used in the extraction process itself. The Promethean nature of fossil fuels is due to the much larger surplus they deliver compared to animate energy converters such as draft animals and human labor. The energy surplus delivered by fossil-fuel technologies is the energetic basis of the Industrial Revolution.[16]

Can solar energy be Prometheus III? The challenge we face is to overcome the constraints imposed by the nature of solar energy and develop it in sufficient quantities to fuel not only the industrialized North, but also the developing South. This is a formidable challenge. There is no guarantee that we will escape economic hardship in the transition from fossil to solar energy, or that current lifestyles can be supported in an all-solar economy. However, great strides are being made in many solar technologies, progress fueled by the growing awareness of the role fossil fuels play in climate change and other pressing environmental problems.

THE OIL INDUSTRY: ISSUES OF CONCERN

6

The Impact of Environmental Concerns on the Future of Oil

Seth Dunn

Concerns about the environmental impact of energy use are nothing new, as evidenced by complaints about air pollution from the burning of sea coal in thirteenth-century England. However, industrialization has greatly magnified this impact and, over the last century, environmental concerns about energy have extended beyond the local to regional and global levels, encompassing not only air pollution but also land degradation, water pollution, acid rain and, most recently, global climate change. While local and regional environmental effects will continue to influence patterns of energy use, this chapter focuses on the impact of growing concerns about climate change on the future of energy in general, and of oil in particular. It is argued that this concern will increasingly drive and accelerate the ongoing "decarbonization" of the energy economy. The conclusion reached is that although the price of oil will continue to be an influencing factor, climate and other environmental concerns, along with new technological developments, may place greater constraints on the future of oil than resource availability.

Mankind's harnessing of energy has long involved the release of carbon atoms, dating back at least as far as the wood fire of the Escale cave in Marseilles, France, over 750,000 years ago. Reliance on wood was a common feature of energy use in most settled parts of the world until the 1800s. Resources of both wood and coal were abundant, but the growing population density and energy use in Great Britain caused wood, which was bulky and awkward to carry, to gradually lose out to coal, which was more concentrated and easily transported. Though it was not noted at the time, this new fuel also happened to use less carbon and more hydrogen per unit of energy, with one or two molecules of carbon per molecule of hydrogen, versus a ten-to-one ratio for wood. Thus began the first wave of decarbonization.

Despite its negative health and environmental effects, coal remained the king of the energy world for the remainder of the nineteenth century and well into the twentieth century. However, the automotive revolution that started in the 1900s eventually favored another new fuel. Oil had an even higher energy density, could be transported through pipelines more easily and emitted less soot. By the middle of the twentieth century oil had surpassed coal as a source of energy. With only one molecule of carbon for two molecules of hydrogen, oil marked the second wave of decarbonization.

At the end of the twentieth century oil was still the world's leading energy source. However, like its predecessors, it faced an emerging challenger. Natural gas, which burned more efficiently, used a distribution network of pipes that made petroleum pipelines appear clumsy and benefited from its designation as the cleanest fossil fuel, began its ascent in the final decades of the second millennium. With one unit of carbon for four units of hydrogen, natural gas represents the third wave of decarbonization.

This molecular perspective on the energy economy is not typical: generally speaking, most experts prefer to discuss energy trends in terms of politics and prices, resources and reserves. Yet the trend toward decarbonization – progressively reducing the amount of carbon produced for a given amount of energy – is as illuminating and important as it is overlooked. Jesse Ausubel of Rockefeller University goes so far as to argue that this pattern lies "at the heart of understanding the evolution of the energy system."[1] Exploring this trend is therefore critical, if we are to understand the future direction of the energy economy.

Indeed, even as the third wave begins its rise, a fourth has appeared on the horizon. Initially building on the natural gas network for its distribution and derived at first from natural gas to run high-efficiency fuel cells, hydrogen, the most abundant element in the universe, could eventually become as dominant an energy carrier as electricity. Eventually using its own full-fledged network and created by the splitting of water from solar, wind and other forms of virtually limitless flows of renewable resources, the hydrogen economy has the potential to free energy from carbon over time.

The first three waves of decarbonization were driven by the quest of growing populations for more abundant, more easily harnessed and better distributed energy sources. Local and regional environmental factors played a limited role in aiding the respective ascents of oil and natural gas. However, the past century's discovery of carbon's role in changing the Earth's climate, the role of humans in adding this carbon to the atmosphere and the potential risks that accompany climatic change have made global environmental concerns a major new factor in the approaching fourth wave.

The Climatic Constraint

The global carbon cycle is one of the most complex and least understood of the planet's large-scale natural processes. An estimated 42 trillion tons of

Figure 6.1
Global Carbon Cycle in GtC (1992–1997)

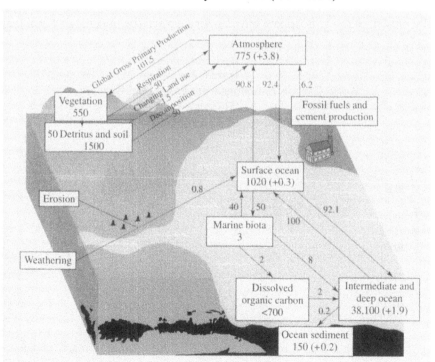

Source: Post 2000

carbon are either housed in or circulate between the atmosphere, oceans and biosphere. The precise amount found in each of these reservoirs is subject to much uncertainty, however, and scientists believe that many of the important fluxes can vary significantly from year to year (Figure 6.1).[2]

One of the most conspicuous characteristics of the carbon cycle is the near-equilibrium of natural fluxes or flows between its various atmospheric, oceanic and terrestrial components. When these flows more or less offset each other, the size of the reservoirs changes little. Beginning in the late eighteenth century, a significant human or "anthropogenic" component has been added to the cycle. Since the dawn of the Industrial Revolution in 1751 and the accompanying large-scale combustion of carbon-based fossil fuels like coal, oil and natural gas, more than 271 billion tons of carbon have been added to the atmospheric reservoir through fossil fuel burning (see Figure 6.2.).[3] With the exception of water, anthropogenic emissions of carbon are the largest single mass flow linked to human activities, ranging between 6.3 and 6.4 billion tons annually.

A clear consequence of this addition to the global carbon cycle has been the elevation of atmospheric levels of carbon dioxide (CO_2), which is naturally present in the atmosphere but also forms when the burning of fossil fuel

Figure 6.2
World Carbon Emissions from Fossil Fuel Burning
(1751–1996)

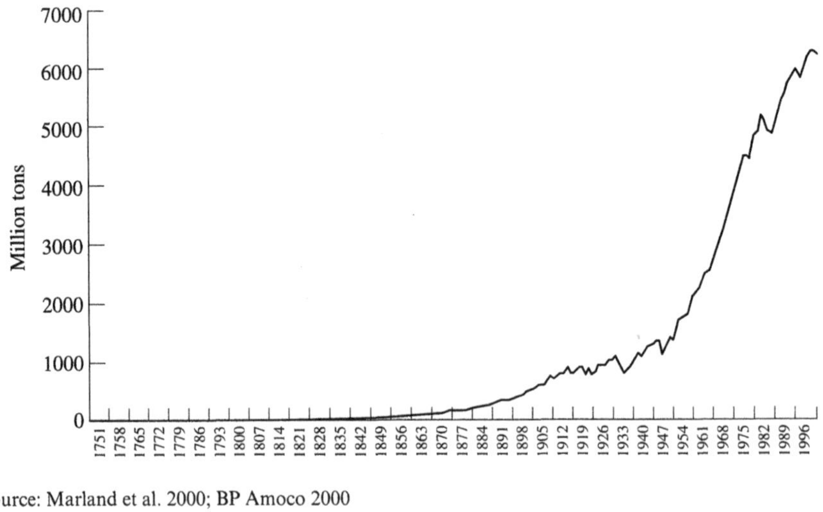

Source: Marland et al. 2000; BP Amoco 2000

releases carbon in the presence of oxygen. Atmospheric CO_2 concentrations reached 368.4 parts per million volume (ppmv) in 1999, marking a 31.6 percent increase from the pre-industrial level of 280 ppmv and a 16.9 percent rise since 1958 (see Figure 6.3).[4] According to samplings of air bubbles trapped in the world's deepest ice core in Vostok, Antarctica, current CO_2 levels are "unprecedented" in relation to the last 420,000 years. Analyses of fossilized plankton suggest that they may be at their highest point in 20 million years.[5]

Carbon dioxide is one of the "greenhouse gases," which also include methane, nitrous oxide and halocarbons, that alter the planet's energy balance by trapping infrared radiation reflected from the Earth's surface, preventing it from escaping to space and causing surface temperatures to rise. The Vostok ice core data suggest a strong correlation between greenhouse-gas concentrations and climate, and indicate that periods of CO_2 build-up have contributed to past global warming transitions between glacial and interglacial periods.[6] Average surface-temperature measurements suggest that another transition is under way. Land-based surveys from NASA's Goddard Institute of Space Studies reveal a temperature increase of 0.6 degrees Celsius since 1866 (see Figure 6.4).[7] Researchers from the University of Massachusetts studying tree-ring samples have concluded that the 1990s were the warmest decade of the millennium, with 1998 being the period's warmest year.[8]

Figure 6.3
Atmospheric Concentrations of Carbon Dioxide
(1958–1999)

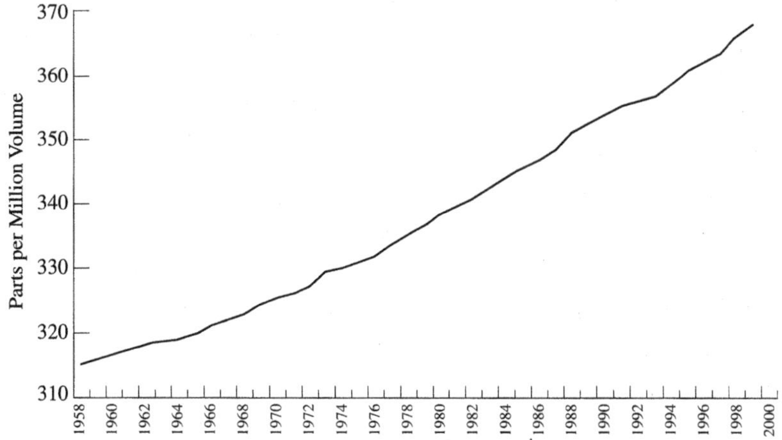

Source: Keeling and Whorf 2000

Figure 6.4
Global Average Temperature at the Earth's Surface
(1866–1999)

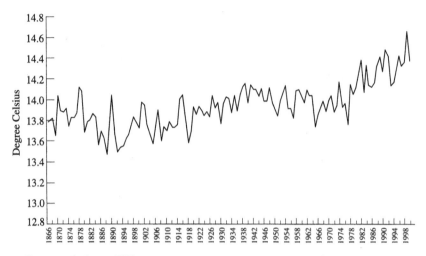

Source: Hansen et al., August 2000

Natural events, such as changes in solar variability and volcanic activity, have also contributed to past temperature changes, such as warming during the medieval period and the Little Ice Age between the seventeenth and nineteenth centuries. However, reconstructions of past climates suggest that human influences have played the dominant role in twentieth-century temperature trends. Thomas Crowley, a geologist at Texas Agricultural and Mechanical University, estimates that natural factors have accounted for only 25 percent of the warming since 1990, with the remainder attributable to increases in greenhouse gas emissions.[9]

The observed surface-temperature rise is actually smaller than that projected by climate models – a "missing warming" is believed to be taking place in oceans, which have warmed dramatically during the last four decades. The oceanic heat storage – half of which, surprisingly, has taken place in deep water – implies that more atmospheric warming lies ahead, as increases in ocean temperature have generally preceded increases in atmospheric temperatures by approximately one decade. James Hansen of Goddard estimates that, because of this ocean-induced delay, surface temperatures will rise 0.5 degrees Celsius over the next century even if greenhouse gas concentrations are stabilized immediately.[10]

The Intergovernmental Panel on Climate Change (IPCC), a UN-designated body of more than 2,000 scientists, has projected a wide range of adverse impacts arising from climate change, including sea-level rise and coastal inundation, more frequent and intense weather extremes, stresses on water and agricultural systems, changing migration patterns and loss of bio-diversity, and greater prevalence of infectious diseases.[11] Perhaps the strongest support for climate models has been evidence of accelerated polar warming and diminishing sea ice and ice sheets in the Northern Hemisphere. Submarine probes suggest that Arctic ice thickness has declined by 42 percent since the 1950s and Norwegian researchers estimate that Arctic summers may be ice-free by 2050.[12] Airborne surveys indicate that the Greenland Ice Sheet is losing 51 cubic kilometers of volume each year. Complete loss of the sheet would raise sea level 7 meters.[13] Such large influxes of freshwater could induce a slowdown of oceanic heat circulation, which has in the past led to abrupt cooling in the North Atlantic region.[14]

The IPCC has developed a new set of scenarios that explore how changes in demographics, socio-economic development and technology may affect the future emissions paths of carbon dioxide and other greenhouse gases.[15] Under the panel's six scenario groupings, carbon emissions from fossil-fuel combustion range from 9 to 12.1 billion tons in 2010, from 11.2 to 23.1 billion tons in 2050 and from 4.3 to 30.3 billion tons in 2100. Cumulative fossil fuel-related carbon dioxide emissions between 1990 and 2100 range from 989 billion to 2.1 trillion tons.

Thomas Wigley of the US National Center for Atmospheric Research has considered the implications of these scenarios.[16] Carbon dioxide levels would range from 558 to 825 ppm, a doubling to tripling of pre-industrial levels. Mean surface temperature would increase by between 1.9 and 2.9 degrees Celsius, three to five times the warming rate of the past century. Average sea level would rise by between 46 and 58 centimeters. The warming is influenced by all greenhouse gases but dominated by CO_2, which accounts for 66 to 74 percent of the warming.

Hansen and colleagues have proposed "an alternative scenario" that combines the reduction of non-CO_2 greenhouse gases and black carbon with success in slowing carbon emissions growth over the next 50 years.[17] They note that these gases have mainly driven the rapid warming of recent decades and suggest that, because their growth rates have declined in the last decade, it may be more practical to focus on reducing them. The authors add, however, that the strong impact of CO_2 on climate remains by far the largest of any single natural or human factor, accounting for just under half of the man-made warming since 1850. They also point out that carbon dioxide will become even more dominant as a greenhouse gas when sulfur dioxide emissions, aerosols that cause a temporary cooling, are reduced. "This interpretation," they assert, "does not alter the desirability of limiting CO_2 emissions."[18]

The Hadley Centre for Climate Prediction and Research, with the UK Meteorological Office, has produced scenarios suggesting that stabilizing CO_2 levels at a doubling of pre-industrial levels (roughly 550 ppmv) would reduce the magnitude of serious impacts in many regions.[19] By contrast, uncontrolled emissions will lead to a 3 degree Celsius temperature increase and 40 centimeter sea level rise by 2080. These conditions will cause substantial tropical forest and grassland dieback in Latin America and Africa, subject an additional 3 billion people to water stress, primarily in Africa, the Middle East and India, and increase the annual number of people flooded from 13 million to 94 million, most of them in southern and southeast Asia. Significant rainforest loss and water resource stress will still result from stabilizing CO_2 at 750 ppmv, but will be avoided by a 550 ppmv stabilization.

Although the objective of the UN Framework Convention on Climate Change is to stabilize atmospheric concentrations of CO_2 and other greenhouse gases at levels that will avoid "dangerous anthropogenic interference with the climate system," scientists have not reached consensus on the stabilization level that would meet this objective.[20] Because a doubling of CO_2 would entail serious dislocations, the IPCC has also considered a more aggressive stabilization target of 450 ppmv. According to the panel, achieving this goal would require cutting carbon emissions by roughly 60–70 percent: to about 2.5 billion tons annually by 2100, and eventually down to less than 2 billion tons per year.[21]

From Solid to Liquid to Gas Resources

The relative shares of different sources of energy have waxed and waned during the last century and a half. In 1850 wood comprised nearly 90 percent of world energy, but its share fell steadily and was surpassed by that of coal in the 1890s, when both fuels accounted for roughly half of the global total.

Coal's piece of the world energy pie expanded to about 60 percent in the 1910s and then shrank, but the fuel remained king until the 1960s, when it was displaced by petroleum as the leading source.[22] Another milestone in the late fossil-fuel era was reached in 1999, with the use of natural gas moving past that of coal for the first time (see Figure 6.5).[23] Today these fuels supply roughly three-quarters of world energy, with respective shares of 32, 22 and 21 percent for oil, natural gas and coal.

This gradual move from solids to liquids to gases has resulted in the use of fuels with lower carbon content. Since 1860 the amount of carbon per unit of energy has decreased by roughly 0.3 percent per year, a relatively slow rate but one that is moving in the right direction.[24] Fuel switching is one factor in the past half century's decline in the carbon intensity of the world economy. Between 1950 and 1999 the number of tons of carbon produced for every million dollars of economic output has been reduced by 39 percent. The average annual rate of reduction during the 1990s was 2.2 percent [25] (see Figure 6.6).

Driving the recent drop in carbon intensity has been a decline in world-wide use of coal, which is the most carbon-intensive fossil fuel and accounts for 36 percent of energy-related carbon emissions. Presently at its lowest point since 1985, coal consumption had an average annual growth rate of

Figure 6.5
World Fossil Fuel Consumption (1950–1999)

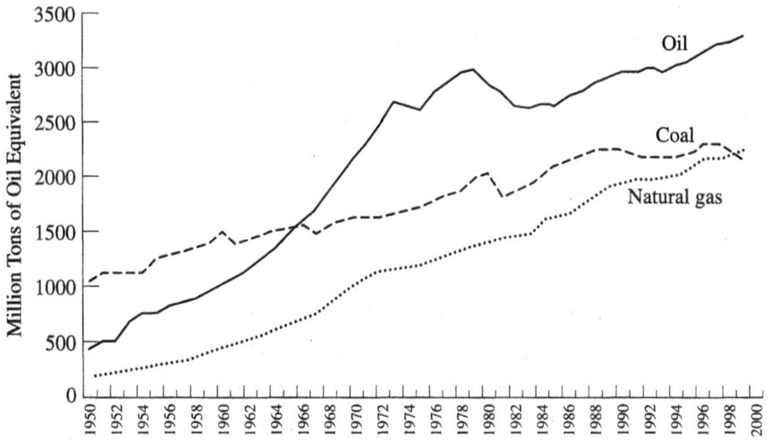

Source: United Nations 1976; BP Amoco

Figure 6.6
Carbon Intensity of the World Economy (1950–1999)

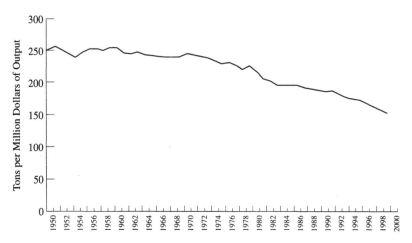

Source: BP Amoco; International Monetary Fund 2000

–0.6 percent in the 1990s. Consumption dropped 7.3 percent just between 1997 and 1999.[26]

Approximately half of global coal consumption takes place in the United States and China. The United States, with a 26 percent share of consumption, saw an 11 percent increase during the past decade. Much of this is due to increased burning of cheap coal in old power plants, some of it illegal, by utilities anticipating industry deregulation. China, which consumes 24 percent of the total world coal resources, experienced a 4 percent drop over the decade, including a 25 percent decline between 1997 and 1999.[27] The decline has been attributed to reductions in coal production subsidies, a steep drop in demand from industry due to closings of inefficient state-owned industries and shifts from coal to natural gas for residential cooking and heating.

Outside the two coal superpowers, consumption trends have varied widely. Coal consumption dropped 24 percent in Europe and 44 percent in the former Eastern Bloc, mostly due to subsidy cuts in the United Kingdom and Russia. In the Asia-Pacific region, India and Japan saw increases of 29 and 17 percent, respectively. South Africa and Australia, which in combination account for 6 percent of world coal consumption, witnessed increases of 15 percent.[28]

Consumption of oil, which contains 23 percent less carbon per unit of energy than coal and accounts for more than 43 percent of energy-related carbon emissions, is still growing, although at lower rates than those of the postwar era. World oil consumption grew by about 1.2 percent on average each year during the 1990s. More than a quarter of global petroleum use occurs in the United States, which boosted use by 11 percent over this period. Japan, with an 8 percent share of oil consumption, increased consumption by 4 percent, while China, with a 6 percent share, boosted consumption by 45 percent.[29]

On a regional basis, the Asia Pacific has experienced the most rapid recent growth in petroleum use, with a 42 percent increase over the past decade. South Korea roughly doubled oil use during this period, while India witnessed a 64 percent increase in consumption.[30] An explosive growth in the use of internal combustion-engine motor vehicles in these countries is driving these jumps, just as steam turbine-powered railroads fueled coal use in an earlier stage of the fossil-energy era.

Natural gas, which accounts for 21 percent of energy-related carbon emissions and releases 28 percent less carbon than oil and 44 percent less carbon than coal, is now the fastest growing of the three fossil fuels. Consumption of natural gas averaged an annual growth rate of 1.9 percent during the 1990s. In industrial nations natural gas has become the fuel of choice in turbines for power generation, replacing coal. The United States, the leading consumer with a 27 percent share of global consumption, expanded consumption by 14 percent in the 1990s. Europe, with a 20 percent share of gas consumption, saw consumption rise by 34 percent. Leading this trend was a so-called "dash for gas" in the United Kingdom, a 75 percent increase enabled by the removal of subsidies to the coal industry.[31]

The most significant declines in natural gas consumption globally have occurred in former Eastern Bloc nations, which consume close to a quarter of the world total but whose economic troubles have dramatically slowed exploitation of the region's reserves. Collectively, these countries saw gas consumption fall 19 percent in the 1990s, with the dominant user, Russia, experiencing a 13 percent drop. Ukraine, another major consumer, saw consumption plummet by 43 percent.[32]

Developing regions have been the major contributors to recent growth in natural gas use, though they still account for only a quarter of the total world consumption. Central and South America, the sites of some of the most recently discovered reserves, have boosted consumption by 58 percent over the last 10 years, with the largest increases occurring in Argentina and Venezuela. The Middle East boosted consumption by 85 percent, with Iran and the United Arab Emirates doubling consumption. In the Asia Pacific, consumption has increased by 70 percent. India and Malaysia have doubled consumption, Taiwan and Thailand have tripled consumption and South Korea has quintupled consumption.[33] Just as in industrial nations, growing reliance on gas turbines for power generation is the main factor that influences this trend, which may in turn be contributing to coal's slowing consumption.

Looking ahead, future policies to reduce carbon emissions may significantly affect fossil-fuel markets. A near-term scenario by the Oxford Institute for Energy Studies sketches the effect of the Kyoto Protocol on fossil fuels, which commits industrial and former Eastern Bloc nations to reducing emissions to 5 percent below 1990 levels between the years 2008 and 2012. The study, *Fossil Fuels in a Changing Climate*, assumes that virtually all countries achieve their targets primarily through domestic action with carbon/energy taxes, voluntary measures from industry and energy-efficiency policies. By 2010 coal production is cut 4.4 percent from its "business-as-usual" trajectory, natural gas by 4 percent and oil by 3 percent. However, given the political uncertainty over implementation of the Protocol, it is difficult to gauge how realistic such a scenario may be.[34]

Fossil-fuel use is also affected in long-term scenarios constructed by the International Institute for Applied Systems Analysis (IIASA) and World Energy Council (WEC) in their book *Global Energy Perspectives*. In all of these scenarios, which use vantage points of 2050 and 2100, "the peak of the fossil era has passed." In two ecologically driven scenarios, the global energy shares of natural gas, oil and coal decline to 11, 6 and 3 percent by 2100, and the overall fossil fuel portion to below 20 percent, which had been their relative share around 1850. Absolute consumption eventually declines for all three, lowering annual carbon emissions to 2 billion tons by 2100 and stabilizing atmospheric CO_2 levels at 450 ppmv. However, this scenario is predicated on "ambitious policy measures" that accelerate energy-efficiency improvements and promote environmentally benign, decentralized energy technologies.[35]

Subsidy removal and carbon taxes are among the possible policy tools to encourage movement toward and eventually beyond lower-carbon fossil fuels. Globally, fossil fuel subsidies, like price supports, favorable tax rates and direct financial aid, total more than $120 billion per year, with the lowest-carbon fuel, natural gas, receiving the least support.[36] These subsidies can be redirected to social programs, such as employee retraining. Similarly, carbon tax revenues can be used to reduce taxes on wages and employment to minimize negative economic impacts. Only six countries – Sweden, Norway, the Netherlands, Denmark, Finland and Italy – have implemented carbon taxes to date.

Improving Energy Intensity

Another element of decarbonization is reducing energy intensity, or the amount of energy required per unit of economic output. Since the Industrial Revolution, energy intensity has annually improved by 1 to 2 percent on average. Energy intensity improvements have been responsible for roughly two-thirds of carbon intensity improvements since 1950, with the remaining third attributable to fuel switching.[37] In the United States one unit of GDP requires less than one-fifth the amount of energy needed two centuries ago. Between 1973 and 1986 energy consumption remained the same while GNP rose over 35 percent, though intensity gains have since slackened.[38]

Energy intensities have generally declined over time due to a variety of factors. These include technical efficiency improvements in power plants and appliances, structural economic shifts to less energy-intensive activities and changing patterns of energy use and lifestyles. However, national circumstances and history are also important. Different consumption and settlement patterns go far in explaining why Australia has a much higher energy and carbon intensity than Japan. The United Kingdom's dash for gas, Japan's leadership in energy-efficiency technologies and the persistence of low fossil-fuel prices in the United States are important factors in these nations' carbon intensities since 1980 (see Figure 6.7).[39]

As with fossil-fuel use, patterns of energy intensity vary regionally. Much of this is due to differing economic structures: as nations move from industrial to "post-industrial" or service economies, energy use tends to rise at a slower rate or to fall. Scientists have also observed a correlation between rising incomes and falling energy intensity, as more efficient technologies become available at higher income levels. The energy-intensity paths of many

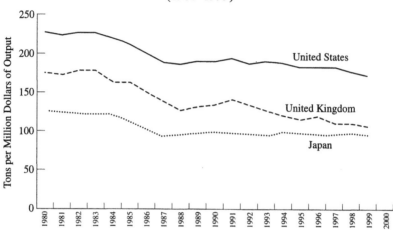

Figure 6.7
Carbon Intensities: Selected Industrial Economies
(1980–1999)

Source: BP Amoco; Maddison 1995; International Monetary Fund 2000

of today's developing countries resemble those of industrial nations at low levels of per capita income. Most of these nations experience a surge in energy intensity as they begin to industrialize, as did China in the 1970s, paralleling India in the 1960s and the United States in the 1900s.

China's subsequent experience demonstrates the opportunity to make major progress regarding energy intensity. As part of the move toward a more market-oriented economy, China enacted an ambitious set of energy-efficiency policies, creating corporations that focused investments on equipment upgrades, efficiency in new construction, building and appliance-efficiency standards and energy conservation centers. As a result, the rates of growth in Chinese energy demand have been cut to half the rate of economic growth. Had Chinese energy intensities remained at 1980 levels, the nation would be emitting one quarter of global carbon emissions, equal to that of the United States, rather than one eighth. Instead, China, India and Brazil now all have carbon intensities that are relatively low for their stage of development and, as one can see through a comparison with Figure 6.7, below the US level (see Figure 6.8).[40]

Although energy use has fallen sharply in former Eastern Bloc nations over the past decade, steeper drops in economic output have pushed the region's carbon intensity upward everywhere except in Kazakhstan (see

Figure 6.8
Carbon Intensities: Selected Developing Economies
(1980–1999)

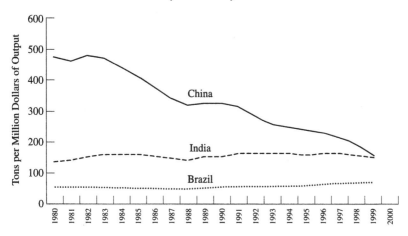

Source: BP Amoco; Maddison 1995; International Monetary Fund 2000

Figure 6.9).[41] As its economy recovers, the region is projected to halve its energy intensity over the next 20 years, although levels will still be 2 to 5 times higher than those in developing and industrial countries. Industrial nations are expected to have average annual improvements of 1.1 percent, compared with 1.3 percent since 1970. A one percent annual improvement is projected for developing nations, though national trends may vary widely.[42] In general, the faster the economic growth and turnover of capital and equipment, the greater the energy intensity improvements will be. Gains are thus expected to be the largest in countries with high intensities that are experiencing rapid economic growth and capital turnover.

Opportunities for deploying energy-efficient technologies in buildings, transportation, and industry are significant.[43] Buildings account for roughly 36 percent of energy use and have lifetimes ranging from five to eight decades, meaning that delays in improving efficiency will lock in waste for many years. It is estimated that improved building designs and appliances available today can cut industrial-nation energy consumption in half. Even greater savings are possible in other regions. Policies that can help overcome barriers to efficiency investments in buildings and appliances include mandatory codes and standards, training, third-party financing, financial incentives and design competitions. Mandatory appliance standards already in

[161]

Figure 6.9
Carbon Intensities: Selected Former Eastern Bloc
Economies (1991–1999)

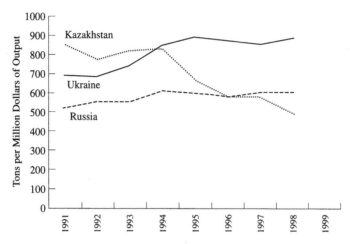

Source: BP Amoco; Maddison 1995; International Monetary Fund 2000

existence in the United States are projected to save more than $160 billion
and 60 million tons of carbon by 2010. For example, the average electricity
use of a refrigerator is one-third what it was in 1974.[44] Mexico, Brazil, South
Korea and China are among those working to develop similar codes and
standards.

One quarter of world energy consumption is for transportation, of which
70 percent is in industrial nations.[45] The failure to strengthen fuel economy
standards has caused efficiency to reach a plateau in the United States, while
most emerging automobile markets use highly inefficient engines.
Considerable potential exists for improving the fuel economy of the existing
internal-combustion model and for introducing far more efficient hybrid-
electric cars. The recent hybrids introduced by Toyota in Japan achieve from
60 to 70 miles per gallon of fuel, compared to the US fleet average of 27.5.[46]
Standards for manufacturers and incentives for buyers can help to bring
these cars on the road, while road pricing and fuel taxes can encourage
greater use of public transport.

Industry consumes close to 40 percent of world energy, with half of this
amount coming from the production of energy-intensive materials such as
steel, chemicals, cement and paper. In transitional and developing economies,
industry consumes more energy than buildings and transport, although

economic difficulties have recently slowed growth in some of these nations. Industrial energy consumption can be cut by one-half or more through more efficient process technologies such as electronic controls, heat-recovery boilers and motor drives, which consume one-half to two-thirds of industrial power in Brazil, China and the United States. Industries that use power on-site rely today on steam turbines with electrical efficiencies of 33 percent, but advanced gas turbines offer 55 percent. By generating power on-site in tandem with the steam and heat used in production, a process known as cogeneration, these turbines, along with newer and smaller devices such as microturbines and Stirling engines, can achieve overall efficiencies of 80 percent and higher. Cogeneration is already a major source of electricity in Europe, providing shares of 30, 34 and 40 percent in the Netherlands, Finland and Denmark respectively.[47]

Structural economic changes also have potential. The gradual move toward less energy-intensive, service-based economies has considerably lowered energy use in industrial nations. But the impact of the information-based or "Internet economy" is hotly debated. While some emphasize the direct effect of greater use of electronic equipment, others argue that the Internet will indirectly temper energy use by reducing the need for energy-intensive manufacturing, retail space and transportation requirements.[48]

Energy analyst Joseph Romm noted that between 1996 and 1999 US energy intensity fell by more than 3 percent per year, compared with average annual rates of less than one percent in the previous decade. One third of this change is structural, due to growth in services and information technology, with the remaining two-thirds due to efficiency improvements. Romm projected that the growing role of e-commerce will lead to US energy intensity gains averaging 1.5 to 2 percent annually over the next ten years.[49] If policies to enhance energy efficiency are added, the improvement could be even greater.

Beyond Fossil Fuels

"New" renewable energy sources are in the position of petroleum about a century ago: accounting for roughly 2 percent of world energy, but gaining footholds in certain regions and niche markets. It was concern about the high price of oil that initially sparked the first modern wave of interest in renewable energy technologies in the 1970s and 1980s. The dynamic growth phase that began in the 1990s, however, features significant technological improvements and is driven in part by policies designed to address carbon emission reduction commitments (see Figure 6.10).[50]

The most spectacular recent growth in renewable energy use has occurred with wind power, which averaged a 24 percent annual increase in the 1990s and now comprises a $4-billion industry.[51] Advances in wind turbine systems have dramatically lowered the generation cost of wind power over the last two decades, to the point where it is becoming cost-competitive with fossil-fuel-fired power generation in some regions. However, its strong market entry also owes much to policy support, particularly electricity "in-feed laws" in Europe that provide generous fixed payments to project developers. Seven of the top ten wind power-using nations are European.[52]

Solar photovoltaics (PVs), which convert sunlight into electricity, have also witnessed significant cost declines and market growth, namely a 17 percent annual average during the past decade. The global solar industry is estimated at $2.2 billion. BP Solar, the leading manufacturer, with a 20 percent share of the world market and estimated annual revenues of $200 million, has its cells in use in more than 150 countries and manufacturing facilities in the United States, Spain, India and Australia.[53] The PV market

Figure 6.10
Global Trends in Energy Use (1990–1999)

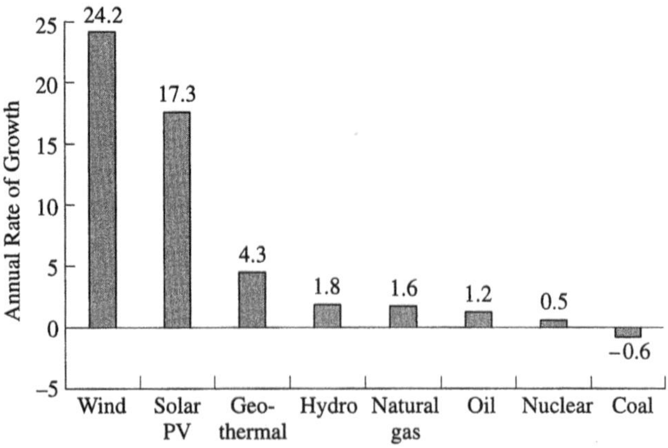

Source: BP Amoco; Worldwatch database; BTM Consult 2000; Maycock 2001; Lund 2000

is experiencing a shift from primarily off-grid uses to grid-connected applications, which comprise the fastest-growing sector due to subsidy programs in Japan, Germany, and the United States.[54] Also poised for growth is the off-grid rural market in developing nations, which is forecast to expand more than five-fold over the next 10 years.

Hydropower, geothermal power and biomass energy have experienced slower but steady growth over the last decade, ranging from 1 to 4 percent annually.[55] Increasingly, new hydropower systems in North America and East Asia are oriented toward small-scale applications. Geothermal power is on the rise in parts of Asia Pacific and Latin America. Use of biomass energy, primarily agricultural and forestry residues, which accounts for as much as 14 percent of world energy and has both residential and commercial uses, is benefiting from modern applications in turbines and factories. Less-established technologies, such as those that harness wave and tidal energy, may yet prove viable.

Several recent publications have explored the possibility of renewable energy providing a significant share of world energy by mid-century. In the Shell Group Planning "Sustained Growth" scenario, renewables first capture niche markets and "by 2020 become fully competitive with conventional energy sources." Solar PV technologies will experience cost reductions similar to those for oil in the 1890s, and between 2020 and 2030 developing countries will turn aggressively to renewable energy.[56]

During the next two decades, in the Shell study forecast, these technologies will become widely commercial as fossil fuels reach a plateau, with wind, biomass and solar PVs achieving market penetration rates similar to those of coal, oil and gas in the past. By 2050 over 50 percent of primary energy supply will come from renewables, with 10 sources each holding a market share of 5–15 percent. The firm responded to its own scenario by establishing in 1997 a Shell Renewables core business, which has earmarked $500 million over five years and has projects under way in solar, biomass and wind energy in Europe, South America, the Middle East, Africa and the Asia Pacific.[57]

Renewable energy is prominent in other studies as well (see Table 6.1).[58] The IIASA/WEC ecologically driven scenarios show renewables reaching a 40-percent share in 2050 and 80 percent in 2100.[59] They note a "changing geography of renewables," as developing nations will take a leadership role in harnessing the resources by the 2020s and will account for two-thirds of renewable energy use by 2050. In *Bending the Curve: Toward Global Sustainability*, the Stockholm Environment Institute (SEI) describes a 25-percent scenario that "requires neither heroic technological assumptions nor economic disruption." It argues that the promotion of renewables may further the goals of economic development and job stimulation; the primary constraints on achieving the energy goal will be institutional and political.[60]

Table 6.1
Scenarios of Renewable Energy Penetration

Scenario	Renewables' Share of World Energy (%)	Date
Shell	50 +	2050
IPCC	50 +	2050
IIASA/WEC	22–37	2050
SEI	25	2050
Roskilde	100	2050

Source: Worldwatch Institute

The SEI study lays out specific policies that can make its scenario a reality. Carbon taxes can be coupled with reductions in other levies, for example in the tax shifts of several European countries. Fossil-fuel subsidies can be phased out. New financing initiatives and economic incentives may spur investment in renewable technologies. Expanded research, development and demonstration can create new technologies. Better information, capacity building and institutional frameworks will overcome barriers to investing in renewables. Global initiatives to transfer technologies and know-how can make these sources the foundation of developing nations' energy economies.[61]

A common question in discussions of the long-term role of renewable energy is whether these resources can conceivably meet worldwide energy requirements. Bent Sorensen at the Roskilde University in Denmark has researched into this question, using an array of economic, population and energy data and projections to create scenarios that consider whether solar, wind, biomass, geothermal and hydropower could meet global energy demand by 2050. His scenarios achieve near-zero carbon emissions and are more expensive than the current system only when environmental costs are neglected.[62]

The Roskilde study concludes that a combination of dispersed and more centralized applications, for example placing solar PVs and fuel cells in buildings and vehicles, and wind turbines adjacent to buildings and on farm-land, plus a number of larger solar arrays, offshore wind parks and hydro installations, will create a robust system capable of meeting the world's entire energy demand. However, the study also stresses that significant additional technological and policy development will be required to realize the scenario.[63]

The policy preconditions for a wholesale shift to renewable energy include a mix of free-market competition and regulation, with environmental taxes correcting marketplace distortions, temporary subsidies to support the market entry of renewables and the removal of hidden subsidies to conventional sources. Taxes will need to be synchronized internationally to avoid differential treatment of energy sources among countries and to be adjusted if the market does not respond enough to the initial price change. In addition, the energy transition will have to be kept on course by continuous setting of goals and monitoring.[64]

The European Union (EU) has already established a target for renewables of 12 percent of total energy sources by 2010 (see Table 6.2).[65] National goals in Germany and Denmark have helped to stimulate rapid wind-power development that has in turn led to more ambitious goals. The Danish firm BTM Consult projects that wind power can supply 10 percent of global electricity by 2020, recommending that nations set specific goals regarding wind power backed up by legally enforced mechanisms such as those now popular in Europe.[66] The United States aims to increase wind power's share of its electricity to 10 percent by 2010, compared to about 0.1 percent today, but has yet to provide policy support.

Table 6.2
Renewable Energy Goals:
Selected Industrial Nations and Regions

Nation/Region	Goal
Denmark	35% of total energy supply by 2030
Germany	25% of electricity from wind by 2010 in the state of Schleswig-Holstein
Japan	3% of total energy supply by 2010
Netherlands	10% of total energy supply by 2020
United Kingdom	10% of electricity by 2010
European Union	12% of energy by 2010

Source: Worldwatch Institute

One of the variables shaping how fast an energy economy based on renewable resources emerges is the extent to which storage systems are developed that can harness the intermittent flows of these sources and store them for later use. Viable energy storage is essential for turning renewables into mainstream sources and engineers have experimented with a long list of candidates, including batteries, flywheels, superconductors, ultracapacitors, pumped hydropower and compressed gas. However, the most versatile energy storage system and the best "energy carrier" remains hydrogen.

Entering the Hydrogen Age

The ultimate step in the decarbonization process is the production and use of pure hydrogen. As noted earlier, the gradual displacement of carbon by hydrogen in energy sources is well under way. Between 1860 and 1990 the ratio of hydrogen to carbon in world energy consumption increased more than six-fold (see Figure 6.11).[67]

Known best for its use as a rocket fuel, hydrogen is produced today primarily from the conversion of natural gas for a variety of industrial applications, such as the production of fertilizers, resins, plastics and solvents. Hydrogen is transported by rail, truck and pipeline and stored in liquid or gaseous form. Though it costs considerably more to produce than petroleum today, the prospect of hydrogen becoming a major carrier of

Figure 6.11
Hydrogen:Carbon Ratio, World Energy Mix (1860–1990)

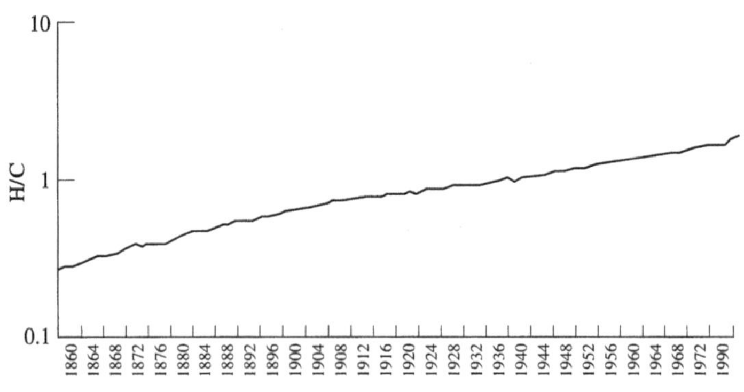

Source: Ausubel 1996

energy has been revived by advances in another space-age technology, namely the fuel cell.[68]

An electrochemical device that combines hydrogen and oxygen to produce electricity and water, the fuel cell was first used widely in the US space program and later in a number of defense applications, such as submarines and jeeps. While these cells were traditionally bulky and expensive, technical advances and size and cost reductions have sparked interest in using them instead of internal combustion engines (ICEs), central power plants and even portable electronics. Their initial costs are several times higher than these conventional systems but are anticipated to drop sharply with mass production.

Fuel cells are on the verge of being used for stationary and transportation purposes. Ballard Power Systems and FuelCell Energy planned to deliver their first commercial 250-kilowatt units in 2001.[69] DaimlerChrysler, which is devoting $1.5 billion to fuel cell efforts over the next several years, aims to sell 20–30 of its fuel cell buses to transport systems in Europe by 2002. It also plans to mass-produce 100,000 fuel cell cars and to begin selling them by 2004.[70] Toyota and Honda have set 2003 as commercialization dates for their fuel cell vehicles.[71]

An important stimulus to the fuel cell market has been the US state of California's requirement that 4 percent of new cars sold in 2003 have zero emission.[72] This requirement has spurred new fuel cell investments and collaborations. The California Fuel Cell Partnership, composed of major car manufacturers, energy companies and government agencies, intends to test 70 fuel cell vehicles by 2003, with energy companies delivering hydrogen and other fuels to refueling stations. The partnership unveiled its headquarters, which include a refueling station and public education center, and first fleet of vehicles in Sacramento in November 2000.[73]

Another region at the vanguard of the hydrogen transition is Iceland, where the Icelandic government and other institutions, like DaimlerChrysler, Shell Hydrogen and Norsk Hydro, launched a $1 million joint venture to create the world's first hydrogen economy in February 1999. The joint venture, Icelandic New Energy, emerged from a parliament-appointed study commission that recommended the initiative. It is now official government policy to promote the increased use of renewable resources to produce hydrogen and geothermal and hydroelectric resources provide 70 percent of the nation's energy. The strategy is to begin with buses, followed by passenger cars and fishing vessels, with the goal of completing the transition between 2030 and 2040.[74]

Hydrogen-powered fuel cell buses are a logical first step, as their lower weight and space requirements and refueling needs can more easily accommodate the low energy density of the gas. Ballard has demonstrated fuel cell transit buses in Vancouver and Chicago, which run on compressed hydrogen gas stored in tanks onboard the vehicles. Hydrogen refueling stations for buses and vans are also appearing in Germany, at Munich airport and in Hamburg, although these will initially supply vehicles with ICEs that use the fuel directly. The Hamburg station intends to eventually import hydrogen from Iceland.[75]

The introduction of fuel cell cars faces three tough technical challenges: integrating small, inexpensive and efficient fuel cells into the vehicles, designing tanks that can store hydrogen onboard and developing a hydrogen refueling infrastructure. The design issue is being overcome by improvements in power density and reduced platinum requirements. The storage issue is being addressed through vehicle efficiency gains, tank redesign and progress in storage technologies such as carbon nanotubes and metal hydrides. Although the direct hydrogen fuel cell vehicle is the simplest and most elegant approach, industry is devoting substantial research to cars that use onboard reformers, which strip hydrogen from gasoline, natural gas or methanol. This may appear preferable to spending the money needed to develop a new refueling infrastructure, but the economics and environmental effects of this approach are less straightforward.[76]

Studies indicate that by the time a critical mass of infrastructure and vehicles are in place, direct hydrogen will be more cost-effective than onboard reformers. Reformer-based fuel cell cars, furthermore, are unlikely to achieve the environmental performance of those using direct hydrogen. The Canada-based Pembina Institute, comparing the "well-to-wheel" greenhouse-gas emissions of various hydrogen vehicle production systems over 1,000 kilometers of travel, found that reforming hydrogen from hydrocarbon fuels does provide an improvement over a gasoline-powered ICE. It also revealed, however, that the fuels most pursued by industry for commercialization in fuel cell cars, namely gasoline and methanol, offered the least improvement, with reductions of 22 to 35 percent. The hydrocarbon resource demonstrating the greatest climate benefits, namely natural gas, whose life-cycle emissions were 68 to 72 percent below those of the gasoline ICE, has been relatively ignored by industry.[77]

One near-term solution to the "chicken-and-egg" infrastructure problem in many countries may be to use small-scale natural gas reformers at fueling stations, relying on existing natural gas pipelines to distribute the fuel. Marc Jensen and Marc Ross of the University of Michigan estimate that 10,000 such stations, covering 10–15 percent of US filling stations, will be enough to motivate vehicle manufacturers to pursue mass production of direct hydrogen fuel cell vehicles. This will require $3–15 billion in capital investment, which "can be weighed against the social and environmental benefits that will be gained as a fleet of hydrogen fueled vehicles grows." Ultimately, hundreds of billions of dollars will need to be invested over decades in a network of underground pipelines engineered specifically for hydrogen.[78]

While natural gas is currently the most common source of hydrogen and the conversion of coal and oil is also being explored, renewable energy sources are likely to eventually produce hydrogen most economically. Electrolysis of water can convert solar and wind energy into hydrogen, and scientists have recently boosted the efficiency of solar-powered hydrogen extraction by 50 percent. Biomass can be gasified to produce the fuel. Other potential renewable hydrogen sources include photolysis, the splitting of water with direct sunlight, and common algae, which produce hydrogen when deprived of sunlight.[79] Basing renewable energy systems on hydrogen will address their inherent intermittency and easily meet energy demand, provided there are continued technical advances and cost reductions in fuel cells and electrolyzers.

Wise decisions made in today's early hydrogen economy can yield enormous economic and environmental benefits. Wrong turns toward an interim infrastructure, on the other hand, may cost millions of dollars in financial assets, lock in fleets of obsolete fuel cell cars and add millions of extra tons of carbon emissions. There is an appropriate role for governments to play in collaborating with transport and energy companies to develop a direct hydrogen infrastructure through greater research into storage technologies and the identification of barriers and strategies. A January 2000 report from the US National Renewable Energy Laboratory concluded: "there are no technical showstoppers to implementing a near-term hydrogen fuel infrastructure for direct hydrogen fuel cell vehicles."[80] However, the study did point out engineering challenges and institutional issues such as the need for codes and standards for hydrogen use.

If governments and industries can produce forward-looking road maps, optimistic scenarios for hydrogen may materialize. In the Oxford Institute's Kyoto scenario, hydrogen will become more competitive due to emissions policies that cause oil prices to rise, undercutting oil supply and reaching production equal to 3.2 million barrels of oil per day by 2010 and 9.5 million barrels by 2020.[81] In their 1999 book *The Long Boom*, former Shell executive Peter Schwartz and colleagues describe a scenario in which fuel cells have displaced the internal combustion engine within two decades, and "by 2050 the world is running on hydrogen, or close enough to call it the Hydrogen Age," with concerns about climatic change a major driver of the transition.[82]

The emergence of a hydrogen economy poses near-term risks for oil-producing nations. In a June 2000 interview with the *Daily Telegraph* of London, former Saudi oil minister Sheikh Yamani attracted attention for predicting that: "Thirty years from now there will be a huge amount of oil – and no buyers. Oil will be left in the ground. The Stone Age came to an end, not because we had a lack of stones, and the oil age will come to an end not because we have a lack of oil."[83] Yamani also expressed his fear that the advent of hydrogen fuel cells, by drastically reducing and eventually eliminating oil consumption, will create serious economic difficulties for countries such as Saudi Arabia. Sharing this concern, OPEC delegates to meetings on international climatic change have argued that they be compensated for the negative impact on their economies from reduced oil demand.

Although concern over climate change and the consequent mitigation policies are indeed likely to undercut the oil market over time, regions that are currently major oil producers and exporters may in fact maintain and even improve their economic situation by broadening their energy portfolios to include natural gas and hydrogen production and export. In its 1996 assessment, the IPCC considered the implications for Middle East oil of a low-emissions scenario in which renewable energy and hydrogen will become prominent energy sources by 2100. Oil exports from the Middle East will then decline absolutely but will grow as a percentage of global oil consumption from 20 percent in 1990 to above 25 percent in 2025 and 33 percent in 2100.[84]

Remarkably, in the IPCC scenario total energy exports from the region will double by 2050 before returning to 1990 levels by 2100. This is because the decline in oil exports is offset by growth in exports of natural gas and hydrogen derived from both natural gas and solar electricity via electrolysis.

"Since H_2 is far more valuable than natural gas and oil," the scenario describes, "the monetary value of Middle East exports increases continually throughout the next century."[85]

One Middle Eastern emirate is already taking tentative steps toward hydrogen, in collaboration with a major automaker. Dubai and BMW are cooperating on a feasibility assessment for hydrogen production, which is expected to recommend additional action along the path to producing hydrogen from solar energy and water.[86] Dubai is interested in hydrogen because it is one of the first emirates in the Gulf where oil reserves are expected to decline, with prevailing expert opinion that oil will last another 10 to 13 years at most.

Dubai is, in turn, of interest to BMW because of its financial strength and its location on the world's "sun belt." In February 2001 BMW launched a six-month world tour of its liquid hydrogen car fleet in Dubai, where H.H. General Sheikh Mohammed Bin Rashid Al Maktoum, Crown Prince of Dubai and UAE Minister of Defense, was given a test drive. In April the United Arab Emirates announced it would devote \$46 billion to environmental research and development, including solar hydrogen, over the next 10 years.[87]

The hydrogen economy is only beginning to emerge. However, as public concern over climatic change grows, energy technologies evolve and the era of cheap oil comes to a close, forward-looking governments and businesses will see the benefits of looking "Beyond Petroleum," to borrow BP's new maxim. While this change will require considerable near-term adjustments, the long-term costs of inaction may be far greater. As former ARCO executive Michael Bowlin told oil industry colleagues at a 1999 conference in Houston, Texas, "We've embarked on the beginning of the last days of the age of oil . . . Conditions are converging for another sea change in the energy use mix – along the spectrum away from carbon and headed toward hydrogen and other forms of energy . . . Embrace the future and recognize the growing demand for a wide array of fuels; or ignore reality and slowly but surely be left behind."[88]

7

New Developments in the Upstream Oil Industry

Mohan Kelkar

At the start of the new millennium, the oil industry can examine its proud accomplishments during the twentieth century and take pride in the fact that it has successfully developed new technologies to exploit new and challenging resources cost effectively. As resources become more scarce and difficult to find, one can expect such efforts to develop new technologies to continue in the future.

This chapter specifically examines the oil industry's accomplishments in the last 10 to 15 years. This period is especially important because the industry has operated in a relatively low price environment. Even in this environment, the industry has developed remarkable new technologies which make it possible to produce hydrocarbons from hostile environments in a cost-effective manner. These technologies include three-dimensional seismic data collection and processing, horizontal and multi-lateral wells, integrated reservoir description and offshore drilling and production. The technologies and their impact on oil production are briefly discussed below.

As the oil companies learned to operate in a new price environment, there was a shift in development of technology. Instead of concentrating on in-house development of technology, oil companies are now more likely to collaborate with other companies, as well as universities. In addition, short-term practical applications are emphasized more than long-term, blue-sky projects. The chapter will examine this new paradigm and its impact on the future development of technologies.

Finally, the chapter will also discuss what lies ahead in the oil industry. This section will specifically evaluate the types of new technology needed for resources that are more difficult to find and produce. These technologies include extension of 3-D seismic to 4-D (time lapse) seismic data collection and processing, sub-sea completion and processing, intelligent well design and control, and effective knowledge management and mining of information.

The challenges of the future are daunting but, if recent history is any guide, the oil industry has the necessary experience, the will and the resources to face these challenges successfully.

Recent Technologies

In this section we will discuss some of the recent technologies that have had the maximum impact on the oil industry. It is difficult to include all technologies that have improved the exploration and exploitation of hydrocarbons. Instead, the goal is to review the important technologies to illustrate how the oil industry has changed in the last 10 to 15 years.

Three-dimensional Seismic Acquisition and Processing

The use of seismic data to understand sub-surface anomalies is not new. It has been used extensively since the 1960s. The principle behind the use of seismic data is relatively simple. Sound signals are sent from the surface to sub-surface formations and the response is measured at the surface over time. Depending on changes in the formation velocities and densities, different responses are observed at the surface. These responses are interpreted and evaluated to build a sub-surface model of the formations. If a particular anomaly is expected to hold hydrocarbons, a well is drilled. In the past seismic data were primarily used for exploration purposes to locate new oil fields.

In the last 10 to 15 years significant progress has been made in utilizing seismic data. The main changes can be summarized as follows:

- *Bigger and Faster Computers*: Seismic data are volume intensive. The processing of seismic data requires large amounts of storage space and computer speed. As computers have become faster, processing and interpretation have become easier. More data can be stored and processed, sometimes in passing, to evaluate the quality of the acquisition and, if necessary, to make modifications in the acquisition strategy. High-speed computers also allow for more frequent collection of data (and hence improvements in data resolution), more complex acquisition geometries (e.g. three-dimensional versus two-dimensional) and more types of data (e.g. multi-component vs. single component). In terms of interpretation, better visualization makes it easier to interpret the data and extract different attributes of information effortlessly.

- *Fast and Efficient Algorithms*: Over the last 15 years better acquisition and processing algorithms have been developed, which make the processing as well as interpretation more efficient.

- *Efficient Acquisition Tools*: The seismic senders have improved so that the source can be better customized for a particular environment. By better managing the frequency range and the strength of the sender, information at the proper resolution and at appropriate depth can be acquired more effectively. Acquisition geometries can be much more complex, which makes the data collection faster and more efficient. The quality of hydrophones or geophones for acquiring different types of data has improved.

- *Interpretation Tools*: In the past seismic data were collected as two-dimensional vertical cross sections. These cross sections were combined to build a three-dimensional image of the reservoir. This type of analysis required a lot of experience and subjective judgment. In modern times, instead of using two-dimensional data and converting it to three-dimensional data, we can directly acquire three-dimensional data. In effect, instead of acquiring two-dimensional cross sections, we can directly acquire three-dimensional cubes. We also have better visualization tools to view these three-dimensional seismic cubes in various ways so that the interpretation of the data is much easier. This allows companies for the first time to examine subtle features. An example would be the many sub-salt discoveries in the Gulf of Mexico. Many of these reservoirs could not be seen using two-dimensional data. With the improvement in three-dimensional seismic data acquisition and processing, many new sub-salt reservoirs are being discovered.

As the quality of the three-dimensional seismic data has improved, so the type of information that can be obtained has also changed. In the past seismic data were primarily used for identifying structural anomalies during the exploration phase. These include unconformities, salt domes and anticlines. Seismic data can now show up stratigraphic changes at sub-surface locations. For example, in a fluvial environment three-dimensional seismic data can identify changes in facies, thus distinguishing between channel sands and shale. With improved vertical resolution, seismic data can also be used for describing reservoir properties such as porosity. This extends the use of seismic data from exploration to exploitation. For example, seismic amplitude and other attributes have been successfully correlated to porosity data. This type of correlation provides us with a tool for describing reservoir properties at inter-well locations where no other data are available. In addition, seismic data may also provide us with a method of monitoring fluid movement. By collecting seismic data at different times and noting the changes in seismic attributes over time, the changes in the saturation of fluids can be estimated and hence the movement of fluid fronts in the reservoirs. This type of evaluation can help the oil companies to manage the reservoirs better and hence optimize oil recovery.

To summarize the improvements in seismic data acquisition and processing, there has been an evolution from two-dimensional slice descriptions of reservoirs to three-dimensional cube descriptions of reservoirs. Acquisition and processing have become much faster and more cost-effective. Instead of the use of seismic data being restricted to the evaluation of main structural anomalies, they can now be used for identifying stratigraphic changes, estimating inter-well rock properties and evaluating fluid movements in a reservoir. Improved seismic data have resulted in higher success rates in discovering oil in more geologically complex environments.

Horizontal and Multilateral Wells

Before the 1990s the traditional approach to accessing a reservoir was to drill a vertical well. In 1989 only 200 horizontal wells were drilled in the world. In 1990 the number of horizontal wells increased to 1,200 (see Figure 7.1). The primary driver for the sudden increase was the realization that a reservoir is produced more effectively by horizontal wells in the Austin Chalk formation in the United States. The formation was fractured and, consequently, by intersecting more vertical fractures, horizontal wells were able to produce the reservoir more effectively than vertical wells. Since then the number of horizontal wells has slowly increased and it is expected that, in the next decade, 50 percent of all wells drilled will be horizontal. As the technology of drilling horizontal wells has evolved, the cost of drilling horizontal wells has substantially decreased. In 1989 the typical cost of drilling a horizontal well was six times higher than that of drilling a vertical well. Today it is in the range of one and a half to two times the cost of a vertical well.

Depending on the specific application, four types of horizontal wells can be drilled (see Figure 7.2). These include:

- *Short Radius:* Most of the re-entry wells are drilled this way. The building rate (going from vertical to horizontal direction) in these wells is very rapid – about one-and-a-half to three degrees per foot. These wells are relatively inexpensive and are normally completed as an open hole. The main disadvantage of these wells is that measurement while drilling (MWD) is not possible. As a result, directional control is more difficult.
- *Medium Radius:* The build rate for these wells is eight to twenty degrees per hundred feet. MWD tools can be used in these wells, making it possible for better directional control. These wells can be either kept as an open hole or can be completed in another way.

Figure 7.1
Number of Horizontal Wells Drilled in the Last Decade

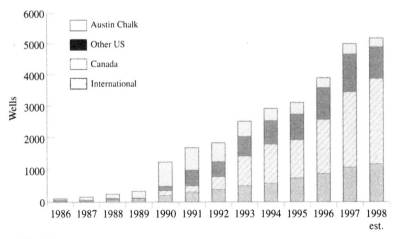

Source: Schlumberger, Inc.

* *Long Radius:* This type of well has relatively low curvature and has a long horizontal section of up to three thousand feet. The build rates are one to six degrees per hundred feet. Traditional MWD tools can be used in these wells, which can be completed or kept as open holes.
* *Extended Reach:* These wells have long horizontal sections to reach the target. They are useful in offshore operations where the operator would like to reach distant horizons from a single platform.

Several reasons exist for drilling horizontal wells instead of vertical wells. These include:

* If the underlying aquifer supports the reservoir, production from vertical wells is subject to water coning. This is due to the difference in pressure between the bottom hole and the aquifer. In contrast, in horizontal wells the pressure is more uniformly distributed in the well, hence the drawdown is smaller and, instead of water coning, water is crested. Because of the low drop in pressure, the water cresting is postponed compared to similar production from a vertical well. This enhances well productivity

Figure 7.2
Different Types of Horizontal Well

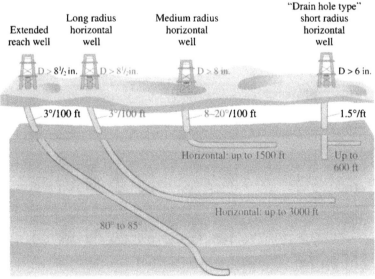

Source: Schlumberger, Inc.

and reduces the operational cost at the surface. The same logic also works when the oil rim is produced in the presence of a gas cap. To minimize gas coning, it is better to use a horizontal rather than a vertical well.

- The vertical well provides limited access to the reservoir. For thin reservoirs, the access is limited to the thickness of the reservoir. In contrast, horizontal wells provide better access to the reservoir, hence improving the productivity of the well.

- Due to improvements in horizontal well technology, horizontal wells can be geologically targeted to optimize reservoir contact. By varying the direction and length of the well, poor areas of the formation are avoided and the best parts of the reservoir are targeted.

- If the reservoir is naturally fractured, horizontal wells can intersect several vertical fractures, hence increasing the effective rate of the formation. For vertical wells, it is difficult to intersect a vertical fracture.

- During water flooding, the use of horizontal wells can make the flooding more effective. By injecting and producing the fluids more uniformly across the horizontal direction, better sweep efficiency is obtained compared to vertical wells. In enhanced oil-recovery techniques, horizontal wells can take advantage of gravity over-ride for improved vertical sweep. For example, when injecting steam or CO_2 in a horizontal well, the injected fluid can move upwards and hence improve vertical sweep. Another horizontal well at the bottom of the formation can act as a drain hole, providing efficient recovery (see Figure 7.3).

- Drilling horizontal wells instead of vertical wells can be environmentally beneficial. By reducing the total number of wells to be drilled from the surface, the drilling footprint at the surface is reduced. For example, a company may be required to drill six vertical wells instead of one horizontal well to produce the same amount of hydrocarbons. This means six drilling locations and their potential environmental effect. By drilling one horizontal well, only one drilling location needs to be created, with less potential for surface damage.

Multilateral wells go one step further than horizontal wells. Instead of a single drain hole, multilateral wells have several drain holes. These wells are especially useful when multiple zones need to be accessed by a single well (see Figure 7.4) or when multiple parts of the reservoir need to be accessed by a single well.

Figure 7.3
Use of Horizontal Wells in Enhanced Oil Recovery

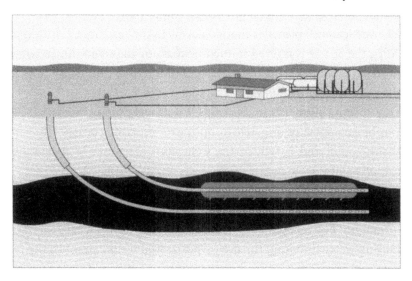

Source: Schlumberger, Inc.

Figure 7.4
Use of Multilaterals for Accessing Different
Portions of the Reservoir

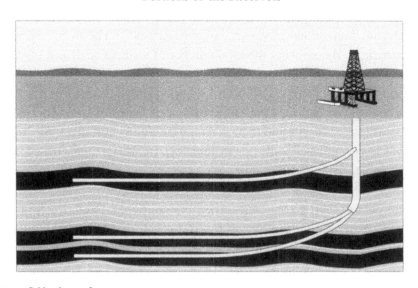

Source: Schlumberger, Inc.

Deep Water Drilling and Production

Oil companies have looked for hydrocarbons in offshore locations since the 1950s. However, the drilling and production activities have been restricted to shallow waters. Even today 35 percent of offshore production comes from reservoirs that are in water depths of nil to a hundred meters, and 44 percent of offshore production comes from reservoirs that are in water depths of one hundred to two hundred meters.

However, as the easy onshore discoveries run out, oil companies are increasingly looking for hydrocarbon resources which are located deep under water. Currently, about one-third of oil production comes from offshore reservoirs. It is expected that this percentage will increase over time. As the oil companies drill deeper and deeper, the technical and environmental challenges faced by them are unique and, therefore, the solutions used are also unique. Some of the technical challenges faced by oil companies, as well as the solutions used, are discussed below.

- *Drilling:* Drilling in deep waters is difficult because it is not easy to anchor a drilling rig in deep water. As a result, the cost of drilling an offshore well is much greater, because of the specifications required for such drilling. New drilling technologies, such as top-drive and dual derricks, have reduced the drilling costs and completion times. Advanced drilling technologies, such as horizontal and multilateral wells, allow operators to drill a limited number of wells to produce the desired horizon more cost effectively.
- *Production:* Once an exploration well is drilled and the economic feasibility of production from these reservoirs has been assessed, the next step is to build a production platform that can be utilized, first, to drill the development wells and, then, to produce from these wells. Starting with fixed platform (FP), which consisted of a jacket (tall vertical section made of tubular steel members supported by piles driven into the seabed), the industry has come a long way. The fixed platform is feasible for about a depth of five hundred meters of water. The new generation platforms are compliant tower (CT), which is a more flexible tower that can withstand large lateral forces. These towers can be used for water depths of between five hundred and one thousand meters. Floating production systems (FPS) consist of a semi-submersible that has drilling

and production equipment. It has wire rope and chain connections to an anchor. Wellheads are on the ocean floor and connected to the surface deck with production risers. These can be used for water depths of two hundred meters to two thousand meters. Tension leg platforms (TLPs) are floating structures connected to the seabed by vertical tensioned tendons. They can be used up to a depth of two thousand meters. In a SPAR platform, the platform is supported by a large vertical structure connected to risers. The platform is anchored to the seabed by tension tendons. Different types of platform are shown in Figure 7.5.

* *Marginal Fields*: Due to the high costs of production platforms, it was previously difficult to exploit small fields economically. However, because of the new generation of extended-reach horizontal and multilateral wells, it has become possible to access these resources and tie them back to the existing production platforms.

* *Project Development*: Oil companies have dramatically reduced the time difference between discovery and actual production from roughly 10 years to about 2 years. This reduction significantly improves the profitability of the project.

* *Low Temperatures*: Typically, the temperatures at the seabed are in the range of 35 to 45 degrees Farenheit. Such low temperatures pose unique

Figure 7.5
Production Platform Schemes

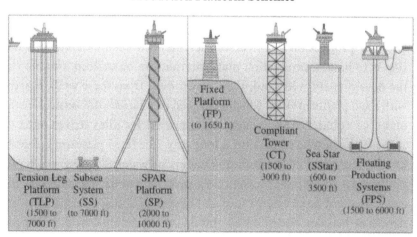

Source: Schlumberger, Inc.

problems. New cementing techniques have been developed to withstand the temperature. New solutions to prevent paraffin wax problems have been proposed and continue to be proposed. Flow assurance has become an important area of research in the last decade.

- *Environmental Considerations*: A recent disaster in offshore Brazil, where a production platform sank due to an explosion, indicates the environmental risks involved in producing from offshore platforms. Oil companies have become increasingly vigilant in protecting the environment while drilling and producing from offshore platforms. Examples of environmental actions by oil companies include using better sonar devices while collecting seismic data so that it does not harm marine life; using specialized drilling mud so that in case of spill it does not adversely affect the marine environment; using sub-surface safety devices so that in case of an emergency production will shut down to avoid a disaster; and metering and disposal of hydrocarbons as soon as they are produced, so that there is no need to store hydrocarbons on offshore platforms, thus avoiding spillage during an emergency.

Integrated Reservoir Management

In the past, oil companies were divided into different groups of people based on their disciplines. Geologists, geophysicists and engineers worked in different groups. The knowledge between the different groups was shared only when necessary. This type of approach did not result in optimal development of a reservoir. Expert, qualitative knowledge was lost in the process, there was mistrust between various disciplines and the transfer of data took much longer because of software incompatibility problems.

As the oil companies have shrunk in manpower, they have learned to use their workforce more efficiently. As part of this process, oil companies have divided people into asset teams rather than disciplinary groups. An asset team will comprise of multi-disciplinary expertise depending on the scope of the project. The team may include a geologist, geophysicist, reservoir engineer, production engineer and petrophysicist. The idea is that by working together on a single asset, communication between different disciplines can be improved, thus optimizing knowledge transparency. The use of an asset management team has improved reservoir management by minimizing the overlap and duplication of work, utilizing and integrating all the available

[185]

data more cost effectively, and using the iterative process of reservoir development, with much feedback, more efficiently.

The human element of the asset management team is further assisted by the development of new technologies, which can integrate various sources of data. Many service companies today can build a "shared-earth model," which integrates various sources of information seamlessly, using various databases. Geostatistics has emerged as a new technology that has the capability to integrate various types of data using efficient algorithms.

In addition to the above-mentioned technologies, various other technologies have been developed in the last 10 years that have had a significant impact on the effective development of hydrocarbon resources. Some of these technologies are:

- *Coiled Tubing*: Unlike pipes, which have to be connected, coiled tubing is continuous. Coiled tubing has been used for slim hole drilling as well as various stimulation and completion activities.
- *Measurement While Drilling (MWD)*: MWD tools have helped in improving drilling efficiency. As more and more complicated wells are drilled, it is important to update the reservoir information as new data are collected. MWD tools measure the sub-surface properties while the well is drilled, thus optimizing the well trajectory continuously.
- *Completion Technology*: Better perforation techniques that allow deeper penetration, and better gravel pack completions that allow a higher rate of completion, are developed for offshore completions.
- *Cased Hole Logging*: New techniques of cased hole logging have allowed better identification of fluid movement and hence bypassed oil.

Technology Development

In the late 1970s and early 1980s, as oil prices soared, oil companies embarked on the most ambitious research activities ever undertaken. Most of the major oil companies developed their own research centers and worked on both fundamental research projects and applied technologies. These research centers were funded by corporate money and were not directly answerable to business units. No technology was considered beyond reach or beyond consideration. Oil companies had their own drilling research, production research, logging research and reservoir-engineering research. The development of technology was emphasized more than the application itself.

As oil prices dipped, the approach to the development of technology changed significantly. Instead of using ever-soaring profits as a way to make money, oil companies developed a new strategy of cutting costs and making their operations more efficient. Among the first casualties of this approach were the research centers and the researchers. The following fundamental changes have taken place over the last 10 years:

- The research centers in most of the oil companies have become technology service centers. The main task of these technology centers is to answer to the business support units. Corporations, except in a few instances, no longer support the technology centers. Instead, the technology centers are required to propose new projects to business units and to obtain support for the projects from these units. This creates a strong tie between the business units and the technology centers and makes the technology centers more accountable.

- Because of the closer tie between business units and technology centers, short-term research projects rather than long-term projects are emphasized. Those projects that may lead to an immediate improvement in operations are favored over long-term, blue-sky projects.

- The overall research budget has been reduced significantly. From $5 billion in 1991, the research budget fell to $2.3 billion in 1999. In the past the oil industry has attracted the brightest minds from various disciplines, because of high salaries and opportunities to work on long-term, academically interesting projects. Now the computer industry has taken the lead as the employer of the best minds.

- Oil companies have decided to concentrate on core competencies and to divest themselves of projects that are not considered core competencies. Oil companies are increasingly outsourcing various types of work to service companies. This means that service companies have assumed the burden of conducting research into various technologies. Today most current technology is developed by service companies rather than oil companies.

- Instead of developing technologies on their own, oil companies have decided to collaborate with other oil companies to pool resources, in order to develop technologies that are considered critically important. A group of oil companies will collaborate with either universities or a service company to develop the product. Several joint industry projects (JIPs) are currently in progress. These projects are of short-term duration and develop specific deliverables within a time constraint.

As an example of successful collaboration between universities and industry, we can briefly examine the various consortia at the University of Tulsa, which has a long history of collaborating with the oil industry, dating back to 1966 when the first successful consortium in drilling was organized. The industry could leverage its money by collaborating with other industry partners on projects that were critical but non-confidential. Since 1966 the University of Tulsa has initiated several consortia. Currently, seven consortia operate on the campus of the university. Although the content of the consortia varies, the principles for all the consortia are the same:

- Solve the problems that are of interest to the industry.
- Leverage industrial support with other governmental support.
- Conduct experimental investigations, which can be scaled to industrial problems.
- Provide tangible deliverables on schedule.

Currently, the total funding for these consortia is in the range of $6 million, making the University of Tulsa one of the best-funded universities in the petroleum-engineering discipline. As an example of the facilities used by the university researchers to produce results, Figure 7.6 shows a loop that is used to investigate paraffin wax deposition in pipelines at various angles.

New Challenges

As the oil industry moves into the new millennium, several new problems need to be solved. Although it is difficult to precisely predict the technologies of the future, one can consider the following as the most likely technologies.

Seismic Data Acquisition and Processing

Seismic data acquisition and processing have made the biggest impact on improving the ability to locate hydrocarbons. Over 36 percent of the oil

Figure 7.6
Paraffin Wax Deposition Loop at the University of
Tulsa Campus

Source: University of Tulsa

companies surveyed by an independent group indicated that seismic evaluation is likely to be the most important technology for the future of the oil industry. Several forthcoming advances will improve seismic acquisition and processing.

In the past oil companies always ran one seismic survey before the first well was drilled. In the time-lapse seismic technique, seismic surveys for the same reservoir are collected over time to detect the differences in the seismic response. By zeroing out the differences where there should not be any change, changes in fluid movements can be detected. An example of this was the Duri steam flood in Indonesia, where, by collecting seismic data over time, the changes in the temperature and hence the movement of the steam could be mapped. As a result, the well configuration could be optimized. A logical next step in the time-lapse seismic method is the continuous monitoring of the seismic response by permanently installing seismic devices such as hydrophones. Seismograms can be installed on the ocean bed to collect such continuous information. Through the interpretation of these responses, the changes in reservoir behavior can be noted and hence a reservoir can be better managed. An additional benefit of continuous monitoring of data is the anticipation of potential disasters and their proactive prevention before they occur. This will result in safer environmental operations.

Typical seismic processing involves collecting single component pressure information and evaluating it to detect various reservoir parameters. However, the pressure wave has several components, which can be detected and interpreted for better understanding of sub-surface characteristics. This requires more storage and increases the cost of interpretation. For complex structures such as sub-salt formations, this increased cost will result in improved detection of oil reservoirs that are hard to find. With increased computer efficiencies and improved hardware, this type of analysis will become much more common.

Another potential application of seismic data is to improve drilling efficiency and do it in a more environmentally prudent manner. The drill bit by itself can be used as a sonic device. By analyzing responses from the drill bit, the drilling path could be changed to avoid environmental disaster. In addition, the path could also be altered to properly locate hydrocarbon resources.

Intelligent Well Solutions

The era of simple vertical drilling is practically over. In a conventional operating well very little information from the well at the sub-surface conditions is gathered. Typically, the wellhead pressure and the production rates of various phases are monitored. Occasionally, a well test is conducted to determine the reservoir information. Instead of collecting occasional information, the future production wells will be able to monitor pressure data on a continuous basis. By collecting large amounts of information, the well performance can be evaluated for a possible change in well configuration and corrective action can be taken to maintain the flow performance. In addition, by monitoring the rate and pressure information from a well, potentially disastrous events can be avoided.

In future, instead of having discrete vertical wells, there will be a network of wells distributed in the reservoir to optimize production. These wells will typically be multilateral, collecting production from various lateral sections and combining them into a single vertical outlet. For optimizing the performance of these types of well, one should be able to control the location of production. For example, if one of the laterals starts producing water, one should have the ability to shut that lateral and still continue to produce from other laterals. If one has remote actuated valves located at different well locations, one will be able to control the flow rates from different lateral sections and maintain the desired flow rate.

To achieve the previous step, one needs to have the ability to measure the flow rates at various portions of the lateral sections. Unlike in production logging, where one needs to insert a spinning device to monitor the flow rate, better flow meters that can be permanently installed in strategic positions will be required. These flow meters will continually measure the flow rates at different locations in the well and send this information to the surface. Using these continually collected data, decisions can be made regarding those laterals that should be producing and those that need to be shut off.

For effective performance of the reservoir, it is critical to monitor the performance continually by sending signals from the well for analysis. To be able to do this cost effectively, improved sensing devices need to be developed that are robust enough to handle the sub-surface conditions for long periods of time. These devices should also be capable of receiving signals so that flow adjustments can be made based on the evaluation of the results. Ideally, these communication devices should be retrievable by wire line so that, if needed, they can be replaced easily.

Some of the continuous monitoring techniques are shown in Figure 7.7.

One of the major advantages of operating intelligent wells is that they allow the management of oilfields in an environmentally safe manner. In many past accidents on offshore platforms, production surged without anticipation, creating an explosion and fire on the platform. By continuously monitoring the well performance, such surprises can be easily avoided.

Improved Visualization

In developing oilfields, geoscientists had to rely on imagination. They cannot see the reservoir but can only imagine how it looks and try to develop the best methods for exploiting it. New computer advances assist them in this effort.

Figure 7.7
Monitoring and Control Devices in a Well

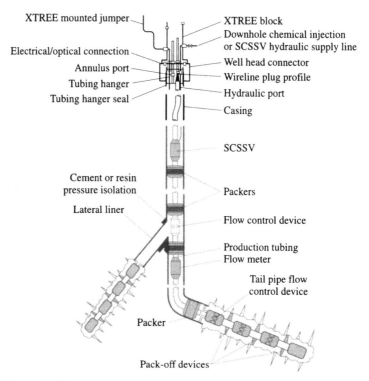

Source: Schlumberger, Inc.

Reservoir description has both qualitative and quantitative components. In three-dimensional space, by visualizing the data, many subtle aspects of the reservoir can be better understood. As three-dimensional seismic data have become more commonplace, it has become increasingly important to be able to visualize this information along with the other data. Three-dimensional images generated by computers are used. However, oil companies have gone beyond visualization on the screen. Instead, several companies have built or are building immersing environments using various sophisticated projection systems, by which the reservoir can be visualized in three-dimensional space as if one is walking inside that reservoir. In essence, one will be able to observe the reservoir on all sides and, by moving and shifting the data around, one will be able to get a better grip on the subtleties in the reservoir description. See Figure 7.8 for an example.

Figure 7.8
ARCO's Cave

Source: Schlumberger, Inc.

An extension of the previous step is to be able to decide on the well path, based on the three-dimensional detailed description. Currently, the development of the well path is based on a limited amount of data. It is much more useful to be in an immersing environment, where the proposed well path can be superimposed on top of the three-dimensional picture; by adjusting the well path, one can make sure that it is intersecting the correct horizons. If, while drilling the well, new information is collected, it may also be possible to generate an updated three-dimensional description, which can be used to redefine the well path quickly.

As new tools are continually developed, the ability to integrate various types of geological, geophysical and engineering data will emerge. It is important that one also has the ability to visualize the three-dimensional description of the reservoir, which is based on various types of information. Such visualization will allow one to assess the value of the new information added to the reservoir description.

[193]

Knowledge Management

Large amounts of information are gathered by the oil companies but are rarely validated, filtered and indexed. Incorrect information, if applied, can result in loss rather than gain in productivity. As oil companies learn to operate with fewer and fewer people, it is critical that the past knowledge gathered from prior experiences be evaluated, analyzed, filtered and stored so that it is easily accessible. Instead of concentrating on data management, oil companies need to emphasize knowledge management, which can be extremely useful to various members of the organization.

Even today each discipline in the oil industry uses its own database. Each database has its own format and is difficult to convert into other formats. Much time is wasted in simply transforming the data into the correct format. For proper integration of manpower as well as knowledge, it is critical that seamless databases be developed which any person in the company can access. These databases should have the flexibility to read information of various types and to convert the information into a suitable format, so that it can be accessed by different software.

With the explosive growth of the Internet, oil company personnel should have ready access to the experts within the company, irrespective of their locations. A system can be developed whereby a company employee can state a given problem and seek the help of an expert in the company. In turn, the database will search various experts to come up with the best expert capable of answering the question. Thus, time would not be unnecessarily wasted in seeking solutions for common problems faced by a particular employee for the first time. This type of solution would have a global reach and can cross any geographical boundaries. This system would also allow the employees of a company to have access to new technologies and gain expertise in those technologies.

Offshore Completions

One of the problems faced by oil companies is the development of marginal oilfields at offshore locations. The prohibitive cost of production platforms in deep water makes it difficult to build a platform for a small field. One possible solution to this problem is the use of extended wells (see Figure 7.9). These extended wells can be drilled with lateral distances exceeding several

Figure 7.9
Current Technology in Extended Wells and Future Challenges

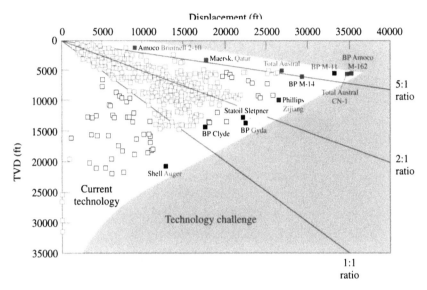

Source: Schlumberger, Inc.

kilometers. As the production from an existing platform declines, the same platform can be used to produce from remote marginal oilfields. The use of an existing platform already built for production from one reservoir can significantly reduce the cost of producing from marginal reservoirs.

In addition to drilling offshore oilfields efficiently, oil companies also need to find ways to produce efficiently from these fields. One option to produce from these reservoirs cost efficiently is not to build a production platform. Instead, the wellheads are located at the sea bottom and the processing platform is built on the seabed. There is no structure on top of the water. All the collection and processing of fluids is done on the seabed and the fluid is transported to a remote location, which could be either onshore or offshore. This type of arrangement is cost effective as well as environmentally friendly, since the companies will not have to worry about de-commissioning of the production platforms (see Figure 7.10).

If the oil companies start using sea-floor completions, one of the key issues that needs to be solved is the flow assurance. Since monitoring of activities can only be done from a distance, oil companies want to make certain that production from the wells is not leaking, that there is no excessive

Figure 7.10
Sub-sea Completion

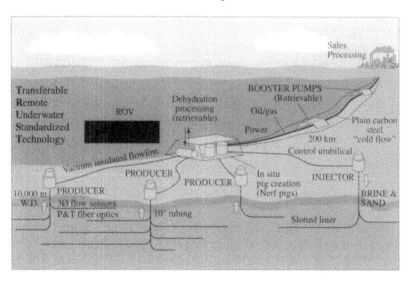

Source: Cliff Redus

pressure drop in the pipelines due to paraffin wax deposition or that there is no flooding of a separator due to high volume slugs. In order to evaluate these problems remotely and address the solutions, proper monitoring devices for flow assurance need to be developed.

Conclusion

The upstream oil industry has come a long way. Some of the important changes that have taken place, or will take place in the years to follow, are listed below:

- The oil industry will strive to find solutions to problems that will have an immediate impact. Gone are the days when oil industry research labs were involved in more esoteric problems that may not have had any immediate applications. The industry has become much more cost conscious and is only interested in solving problems which can have a near-term impact. The service companies and the research universities will play an increasingly important role in developing new technologies in the oil industry. Oil companies will collaborate on critical but non-confidential research ideas to reduce costs and increase the development pace of new technologies.

- The exploration of new oilfields will become more difficult as we look for hydrocarbons in deeper formations and more hostile environments. One of the remarkable aspects of the oil industry is that it has always developed the necessary capacity to surmount new challenges. This trend will continue in future. As the world demand for oil increases, oil companies will continue to add new reserves so as to ensure an uninterrupted supply. At least in the next 50 years, discovering new oilfields and efficiently producing from existing resources will meet the world demand for oil.

- With the advent of computer technologies, oil companies will increasingly rely on computers to manage oilfields. Through continuous monitoring of oilfields, oil companies will be able to better exploit existing and new oilfields. Computer technology will also allow employees better access to new technologies and help them make quicker and better decisions.

- The oil companies have become much more environmentally conscious. The self-imposed environmental audits by many major oil companies are an excellent example of such awareness. Many new technologies are specifically developed to minimize or avoid environmentally disastrous consequences.

The author would like to acknowledge several people who helped with the completion of this chapter. Schlumberger Inc. provided several presentations and written documents which were extremely helpful. Special thanks are due to Mr. Mehmut Sengul from Schlumberger. Several faculty members from the University of Tulsa provided background material. Specifically, the author acknowledges the assistance of Dr. James Brill, Dr. Chris Liner, Dr. Stefan Miska and Dr. Ovadia Shoham. In addition, the author would like to thank Dr. Cliff Redus, who provided information about sub-sea completions.

8

Seeking Stability in the Oil Market

Michael C. Lynch

The history of the petroleum business has been one of volatility and efforts to overcome that volatility. The early market in the 1870s included a very active futures market,[1] unregulated and yet fairly efficient, reacting primarily to the news of new fields being discovered or old fields going dry. Both happened often and unpredictably enough to contribute to severe boom and bust cycles in the industry. Additionally, because of the low barriers to entry, the industry tended to overbuild during boom years, worsening the down-cycles.

For this reason, John Rockefeller created the Standard Oil Trust, controlling enough of the market to take the position of price maker, while buying and sometimes closing competing operations, including refineries, to reduce the amount of surplus capacity. He was successful in that endeavor and the prices were much more stable from 1890 to 1910, at the peak of Standard Oil's power, than before or after (see Figure 8.1). Of course, there was no world market as such at this time and the US provided 56 percent of world oil production over this 20-year period.

As the great international fields came on-line in the first part of the century and the industry faced new competition, both from the successors of the dissolved Standard Oil and the two new major actors, British Petroleum (BP) and Royal Dutch Shell, price wars began to become common.[2] Additionally, the Great Depression and the discoveries of the supergiant

East Texas fields caused a collapse in the US market in the early 1930s and sent the American industry, dominated by small producers, in search of price stability (see Figure 8.2).

Figure 8.1
US Wellhead Prices (1860–1910)

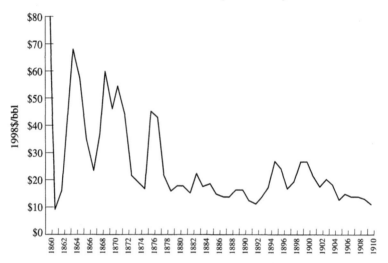

Source: *Energy Price Statistics*

Figure 8.2
US Wellhead Prices (1910–1950)

Source: *Energy Price Statistics*

The response came in the form of production quotas set by the Texas Railroad Commission and emulated by other oil-producing states in the USA, which used estimates of supply and demand and enforced its decisions through the police powers of the state. Since the United States produced up to 52 percent of the world's oil as late as 1950, this was an important factor in creating global price stability.

Of course, the marginal barrel determines the market price and, given that the average US well was producing as few as 12 barrels per day by 1950, compared to 10,000 b/d in the gushers of the Middle East, only by controlling the expansion of production there (and later North Africa) could the market be restrained. To that end, the combination of the Achnacarry agreement and the interlocking concessions in the Middle East were exceptionally effective.

At Achnacarry the major oil companies essentially agreed not to overproduce in an effort to prevent prices from sliding. Given an estimate of annual demand growth, companies expanded capacity and production to meet it. Since the major companies had interlocking concessions in the Middle East, each one's production plans were generally known and it was not feasible to cheat in a major way. Again, the structure of those concessions was such that any member who overlifted above the desires of the other partners would be penalized. This discouraged them from producing too much or seeking to expand their market share, and prices were relatively stable.[3]

However, during the 1960s the surplus capacity in the US gradually disappeared as drilling fell off, and the independents increasingly challenged the majors overseas, breaking into new markets and adding to a growing oil surplus that kept prices under pressure (see Figure 8.3). While data on the price of internationally traded crude during this period is difficult to obtain, the evidence suggests that prices declined about 40 percent from 1960 to 1970. Most of this was absorbed by company margins, which had been quite high, but the concessionaires had cut their royalty payments to the producing governments in 1959 and tried again in 1960. These cuts were primary stimulants to the formation of OPEC.

The 1970s saw two politically motivated price spikes, the first when the OAPEC nations cut production in support of the second Arab oil embargo in 1973, and the second when the Iranian Revolution reduced output from that country for a brief period. Market behavior during such periods is best dealt with separately, since it does not reflect reactions to normal economic stimuli.

Figure 8.3
US Average Wellhead Price

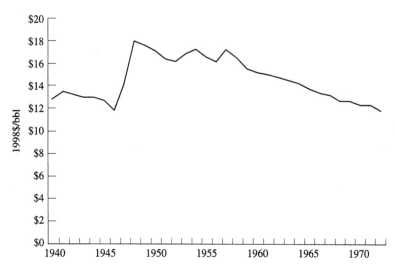

Source: *Energy Price Statistics*

Figure 8.4
US Refiners' Acquisition Cost for Imported Crude
(1974–1985)

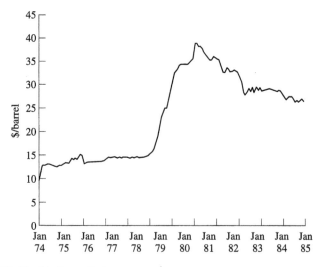

Source: USDOE, *Monthly Energy Review*

Figure 8.5
OPEC Market Share

Source: USDOE, *Annual Energy Review*

From 1974 to 1978 OPEC faced an easy task. Revenues were high, so the pressure to increase production had abated somewhat, and some members decided to conserve resources and cut production. However, the second price spike in 1979/80 reduced the demand for OPEC oil so much that the organization had to respond aggressively to keep the price from collapsing. Saudi Arabia took over the role of swing producer, letting its production vary according to supply and demand fluctuations at the price that OPEC set.

Due to this, world oil prices were generally stable from 1974 to 1985, shifting very little from day to day or month to month, the Iranian oil crisis being the exception, as prices soared in response to the disruption (Figure 8.4). However, the collapse in OPEC market share (see Figure 8.5) – due to weak demand and higher non-OPEC supply – reduced its power greatly and Saudi Arabia found itself heading toward an end to its exports (see below). As a result, Saudi Arabia acted to bring the price down to a new, lower level and also refused to act as the swing producer, instead insisting that all OPEC members cooperate and set formal production quotas.

Since that time, prices have been much more volatile and the past two years have been beyond anything seen during a normal market period in the past three decades (see Figure 8.6). This should not have been very surprising. This author noted that the decline in oil prices and a resumption of oil

Figure 8.6
US Refiners' Acquisition Cost for Imported Crude
(1974–1998)

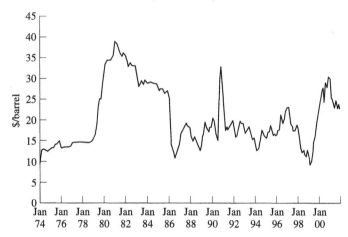

Source: USDOE, *Monthly Energy Review*

demand growth, which were correlated, should eventually lead to a decline in the enormous amount of surplus capacity, which was troubling not only OPEC but the tanker and refinery markets, and that this would lead to greater price volatility.[4]

Understanding the Short-term Oil Market

It is not hard to find evidence of inaccuracy in short-term oil market forecasting. In early 1998 the major magazines and journals typically reported expectations of stable prices for 1998, despite rapidly weakening oil markets.[5] Then, during 1998, many oil market analysts began to argue that oil prices would remain low, in the $12–14/bbl range for WTI, for several years, and in early 1999 a survey in one oil newsletter found that nearly all forecasters had lowered their forecasts for 1999 oil prices a mere six weeks before they began to jump (see Figure 8.7).

Although most economists would argue that short-term price volatility is a fact of life for commodities and that they are irrelevant for long-term trends, the layoffs and dislocation in the oil industry in 1999 tend to suggest that the question remains important.

[204]

Figure 8.7
Forecasts of WTI for 1999

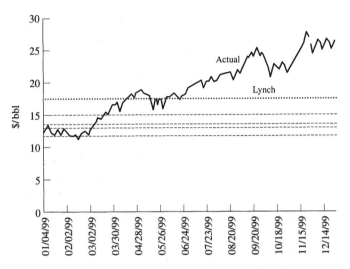

Source: *Petroleum Intelligence Weekly*, Jan. 21, 1999; actual figures from *Wall Street Journal*

Figure 8.8
The Inevitability of the 1985/86 Price Collapse

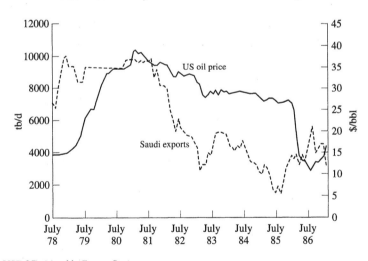

Source: USDOE, *Monthly Energy Review*

[205]

The 1985/86 Oil Price Collapse

There is certainly ample indication that some failures to predict short-term oil prices are more the result of bad analysis, inattention, bias or other problems. For example, in the early 1980s Saudi Arabia acted as the swing producer in OPEC, setting prices and letting production fluctuate according to demand for its oil. This helped to keep short-term oil prices relatively stable, especially compared to more recent behavior. However, Saudi Arabia was bearing nearly the entire burden because, with rising non-OPEC production and increasing pressure on OPEC, its production was swinging down a lot more than it was swinging up and, as Figure 8.8 shows, its monthly export level (taking out consumption) was dropping sharply to the point where, by mid-1985, its exports had reached only 2 mb/d, most of which was needed for barter deals for airliners and so forth.

Clearly, Saudi Arabia had reached the physical limit of its ability to stabilize prices by 1985 and something had to be done to increase its cash sales. Only two options were available: to make prices drop so as to raise demand for OPEC oil overall, and/or to force other OPEC members to cut back and give the Saudis a greater share of OPEC production. Since the latter would almost certainly require that Saudi Arabia initiate a price war, it is clear that oil prices should have been expected to drop in 1986. Yet few anticipated this. This can be seen by the way in which US drilling levels dropped only as the price actually fell, as Figure 8.9 shows.

Petroleum Intelligence Weekly, on November 25, 1985, still reported: "The surge in spot oil prices is persisting as refiners search but fail to find any signs of excess oil on the market, despite repeated warnings of an imminent flood of supply."[6]

Indeed, the late-1985 market behaved almost irrationally, as speculators, recalling the sharp increase in heating oil prices in the 1984–85 winter, went long on the US market, driving up heating oil and crude prices (see Figure 8.10), even after Saudi Arabia announced in September that it was abandoning production restraint and seeking to increase its market share. This would seem to be a major case where the spot and futures markets were driven by speculative fever, not by realistic expectations or the underlying market fundamentals.

Figure 8.9
The US Upstream Industry's Failure
to Anticipate the Price Collapse

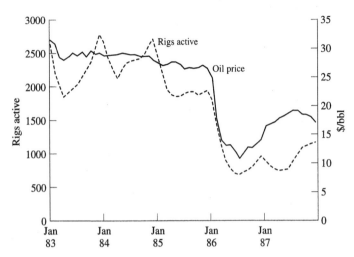

Source: USDOE, *Monthly Energy Review*

Figure 8.10
Heating Oil Prices and the 1986 Oil Price Collapse

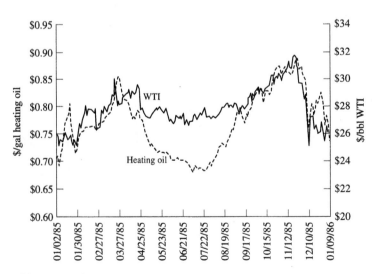

Source: *Wall Street Journal*

Why is it so hard?

In point of fact, it is much harder to get short-term oil prices right than to get them wrong. Indeed, it is arguably harder to predict next year's oil price than to predict the average price over 10 years.[7] Partly, this reflects the fact that a long-term forecast should not refer to an immediate price, which can be very volatile, but the average over a period. Getting the general trend correct but missing a short-term peak or trough (as in 1996 or 1998) is thus acceptable in a long-term forecast, but represents a significant error in a short-term forecast.

The uncertainty levels for the short-term are also very high. Data lags for production, consumption and inventories can be 3–6 months in the industrialized world and longer in the developing world, which is of increasing importance.[8] The recent controversy over missing barrels largely reflects data discrepancies. In Figure 8.11 the missing barrels are shown. These are the estimates from *Oil Market Intelligence*, a leading trade journal, and they include both the initial estimates, which are published in the third week of the subsequent month and the revisions as published five months after the actual period, the last published estimates. As can be seen, the amount of oil that is not accounted for is enormous and fluctuates quite a bit.

Figure 8.11
Missing Barrels

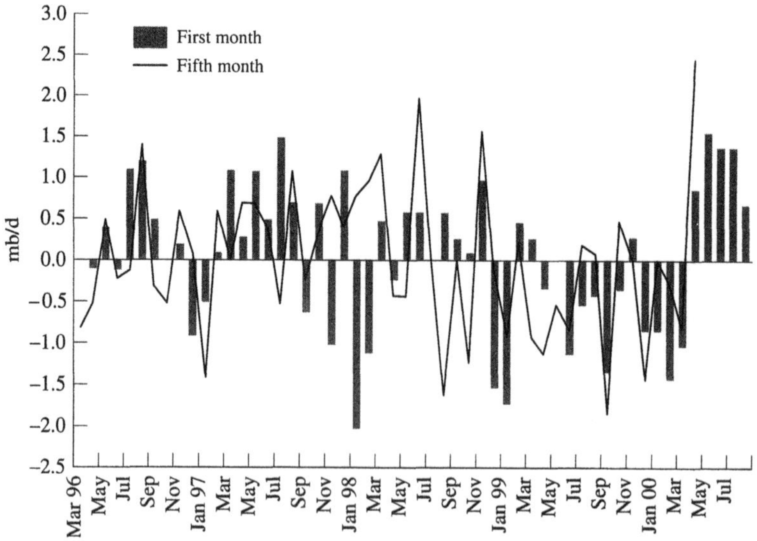

Source: *Oil Market Intelligence*, various issues

Figure 8.12
Heating Degree Days

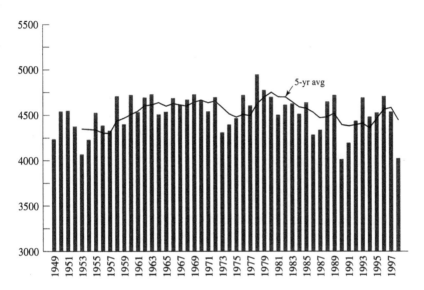

Source: USDOE, *Monthly Energy Review*

These uncertainties are much more important for the short-term than the long-term, since they represent the current trends that will be shaping the immediate future. A lag of six months for consumption data, for example, represents a minor uncertainty when forecasting 10 years into the future, but a major one for a one-year forecast. Revisions to data are often quite large, even a year later, enough so that a statistical glut can completely disappear at a later date. Thus, the uncertainty in the short-term, especially in times of economic transition such as recession or recovery, can be much higher than for the long-term.

Also, there are a number of factors that are inherently unpredictable in the short-term. Weather is a good example. Although some think that the average weather for a decade is uncertain, the possibility of an abnormally hot or cold winter or summer affecting oil demand significantly is much greater than that the long-term trend will suddenly deviate from historical experience. As Figure 8.12 shows, US heating degree days have shifted only slightly over the past 50 years. The average heating degree days in the 1960s were only 1 percent different from those of the 1970s and 1980s, but the average annual changes were 3 percent.

Figure 8.13
OECD Oil Consumption

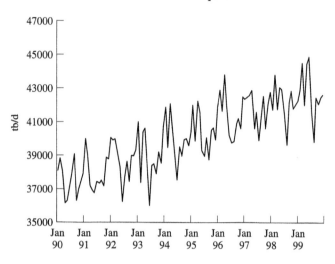

Source: USDOE, *Monthly Energy Review*

Figure 8.14
US Real GDP Growth (Quarterly, Seasonally Adjusted)

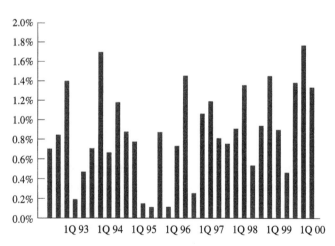

Source: USDOE, *Monthly Energy Review*

Indeed, there have been occasions when short-term weather caused short-term prices to move substantially, the classic case being the 1992 hurricane, which disrupted US natural gas production and sent prices soaring. On occasion, cold winters have caused heating oil and crude prices to rise sharply, while warm winters have caused the reverse to happen.

Other variables have a similar effect, including economic growth, as well as the lumpiness of some oil production. As Figure 8.13 shows, there is a clear seasonal component to OECD oil consumption, and yet there are also other factors at work, to such a degree that there can be a fluctuation of 2–3 mb/d in a month. Indeed, even economic growth varies enormously and often unpredictably in the short-term (see Figure 8.14).

OPEC's Role in Setting Price Stability

However, in the past decade and a half OPEC policy has been the strongest, although not the only, determinant of oil prices in the short-term, along with perceptions of this policy. Although managing an oil grouping comprising diverse nations is a complex task, OPEC has recently convincingly demonstrated that its demise was greatly exaggerated. OPEC's ability both to manage a soft landing and to determine where the landing is remains a question which requires an understanding of not just the supply/demand balance but also likely OPEC behavior. Knowing what OPEC will do over the longer term is of course predicated on what OPEC thinks it should do or wants to do, which are not necessarily the same thing.

Price Stability and Cartels

One of the most amazing things in the debate about oil prices over the past 25 years is the extent to which cartels have acquired a pejorative connotation, and the degree to which most people seem ignorant of the larger field of commodity price stabilization.[9] For one thing, cartels are actually quite common and have been used effectively around the world. This is especially true of what are known as "recession cartels," where companies are allowed to share a market in an organized way during times of economic downturn. The thinking is that companies should not be penalized for macroeconomic events beyond their control.

However, in addition cartels are a subcategory of commodity price stabilization agreements, something that has been quite common in the world generally, particularly since the Depression and World War II. At the international level, a number of commodities, including copper and coffee, were governed by price stabilization agreements, which were at least aimed at providing economic aid in the form of export stability to less developed countries that were reliant on particular commodities. However, most of these have broken down due to a combination of mismanagement and politics, including a growing conviction that they are economically inefficient by nature.

Many industrialized nations use a wide variety of programs to stabilize and especially raise agricultural product prices (and coal in Germany). Many have been in place since the collapse of these prices during the Great Depression, and most have been roundly denounced as horribly inefficient, but reforming them has proven extremely difficult.[10] The US, which often preaches the virtues of unfettered markets, provides less support for its farmers than many of the other OECD countries, but nonetheless it too has a broad variety of programs.

In fact, wages were historically much more volatile than they are now, and early in the century some economists argued that this was beneficial and denounced efforts by unions to prevent wage cutting during periods of economic weakness.

Petroleum Economics

Amazingly, every price drop still brings to the fore those who perceive the end of OPEC, and every market recovery is taken as evidence that OPEC has "learned their lesson" and firmly regained the driver's seat. No one thinks that any given cycle is just another cycle or even recalls previous cycles, much less explains its irrelevance to current conditions. Thus, it can be argued that OPEC's role in the market has changed very little in the past decade and a half.

The Nature of OPEC

OPEC's success lies not in its organizational structure, but rather despite it. The grouping itself lacks decision-making powers. It merely provides information-gathering, analytical and administrative support for the ministers from the member countries, who make all the important decisions. OPEC itself has no enforcement powers, or even audit powers, which means not only that it cannot make the members do what it wants, but it cannot officially find out if they are complying. This is not the way to design an effective organization, either in theory or in practice.

Why then has it been successful? It is because the stakes are so large. For all of the complaints from member countries about the burden of production quotas, they all realize that in a weak market a production cut of 10 percent can yield a doubling of prices. Such a simple action can be worth $75 billion a year, more than enough to make ministers forget arguments about which year to base quotas on, the definition of crude oil, their relative social needs or indebtedness, need for disaster relief and so forth, at least for a while.

However, this implies that OPEC should find it easy to maintain cohesion and keep prices approximately stable at the optimal level, and 1998 showed that this is definitely not the case. In reality, members have every incentive to cheat, just as non-member producers have every incentive to refuse to join. Indeed, game theory has suggested that it is valuable to cheat as long as punishment is neither sure nor rapid.

Before discussing what OPEC *will* do, it is important to think about what OPEC *should* do and what it *can* do. OPEC wants to optimize its revenue, which in theory equals the sustainable price level. In practice, though, these are not the same, partly because it is not clear what either one is, but also because of the internal dynamics of the organization.

What OPEC Can Do: Market Power

OPEC market power comes from four sources: first, its market share, which determines the relative importance of price changes on its demand levels; second, its financial situation, which indicates its ability to sacrifice short-term revenue for longer-term gain; third, its political cohesion, or the willingness of the members to work together; and fourth, the market balance, including inventory levels and the amount and distribution of surplus capacity in OPEC, which indicates the physical ability of members to cheat.

Right now, most of these factors are favorable, as OPEC's market share has increased from the low of 30 percent in 1985 to 43 percent now, although the improvement has been slow in recent years. Although it is always difficult to predict factors like OPEC market share, it seems more likely to increase, if only slowly, in the future.

Its current poor financial situation is both a hindrance and an advantage. Revenues have dropped 50 percent, even from the pre-1979 level (assuming 2000 is a short-term deviation), and mismanagement and "Dutch disease" have left most members constrained in their ability to manipulate short-term export levels to support price-management strategies. However, at the same time, precisely because they are in such dire straits, they are less able to afford price wars than in 1998/99 and should be quicker to respond to severe price declines.

The importance of political cohesion (which is a euphemism for a variety of interactions among the member states) is best illustrated by the way the change in government in Venezuela brought in new policy makers more willing to accept production quotas. Unfortunately, this is hard to quantify or predict, since it would require forecasting the identity of political leaders and/or their politics, as well as their ability to learn.

The role of surplus capacity is probably best demonstrated by the manner in which this year's quota negotiations have been influenced by the number of members who prefer small increases, because they are unable to raise sales more than a few percent. For example, only Algeria, Kuwait and Saudi Arabia are operating below 90 percent capacity utilization. While capacity utilization is likely to abate over the next year or so, it is unlikely to ever drop as low as it did during the 1980s (Figure 8.15).

Given the current inventory levels and the moderate amount of surplus capacity, the implication is that OPEC at present has significant market power. Of course, this is at least in part a short-term phenomenon, but it still implies that it has considerable freedom to decide the price it wants.

What OPEC Should Do: Optimization

Advising OPEC on long-term strategy is a rather daunting task, since it has been done often and nearly always badly. Since the first price shock in 1973, there has been a great consensus that OPEC's market share and power would inevitably strengthen. OPEC has always been told that its problem will be to fend off desperate consumers, stabilizing prices from the high side. Figure 8.16 shows some of the IEA forecasts of non-OPEC production; they are

Figure 8.15
OPEC Surplus Capacity

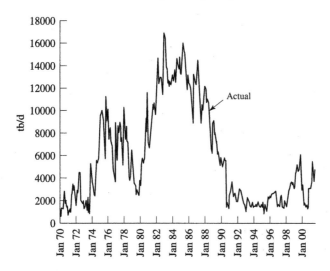

Source: Central Intelligence Agency, 1970–75; *Petroleum Intelligence Weekly* and *Oil Market Intelligence* from 1975

Figure 8.16
IEA Forecasts of Non-OPEC Crude Oil Production

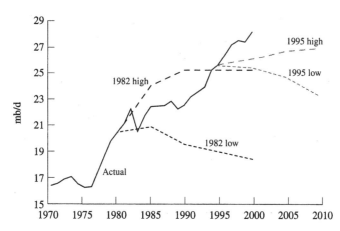

Source: 1982 forecasts from IEA, *World Energy Outlook*, 1982; 1995 figures from IEA, *Oil, Gas and Coal Supply Outlook*; actual from BP Amoco, various years

hardly alone in their historical tendency toward excessive optimism about the prospects for OPEC. Yet throughout this period OPEC has repeatedly struggled to regain lost market share and faced repeated periods of price weakness, even collapse. OPEC must be starting to think that the value of listening to experts lies in learning what *not* to do.

Without question, OPEC members love the current high prices. However, this is a far cry from determining what price will provide the maximum net present value (NPV, which we'll define as optimal). Not only are supply and demand elasticities still not well known, but they are also clearly changing over time to a degree that is poorly understood and possibly unpredictable. Also, what would be optimal for a homogenous grouping differs from what is optimal for the diverse members that make up OPEC, with their widely varied economic and resource situations.

Ultimately, if it is all but impossible to discern what the optimal price level would be, even making Herculean assumptions and simplifications, it should be no wonder that OPEC's strategy appears to be more responsive to the daily price levels than some coherent long-term plan.

Sustainable versus Optimal Price Levels

Beyond the question of what price would be optimal for OPEC, it can be argued that there are different considerations that determine whether a price is sustainable. Of course, the optimal price should take into account competition from other fuels and input factors (supply/demand effects), but this is not the same as the dynamics of the organization, which, as long as OPEC has no central authority and consists of members with different goals and attributes, equates to a very large omission in the calculation.

The 1986 price collapse is illustrative. It is true that demand for OPEC oil dropped by 14.5 mb/d in the early 1980s, but this largely occurred before 1982. From 1982 to 1985 OPEC only needed to cut production by 5 percent per year to stabilize the market, hardly an enormous burden (Figure 8.17). However, the task fell almost completely on Saudi Arabia, which by 1985 was almost a net importer, as was shown earlier in Figure 8.8. Saudi Arabia not only forced other member countries to cut production, but also insisted that OPEC accept a lower price ($18) as more "sustainable," that is, one that would increase demand for its oil and thereby make it easier for the organization to allocate quotas. The result was 12 years without a major collapse or price war.

Figure 8.17
Change in OPEC Production

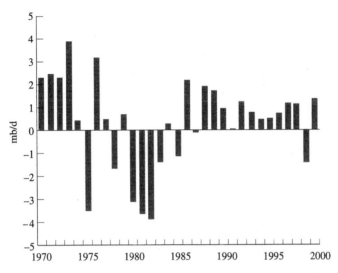

Source: USDOE, *Annual Energy Review* and BP Amoco. Data for 2000 from IEA, *Oil Market Report* and estimates by the author

It is true that the apparent success is exaggerated somewhat by the Gulf War. Growth in OPEC sales of 1 mb/d/yr from 1985 to 1990 made it much easier for OPEC to maintain cohesion, but by 1990 aggressive expansion in Iraq and Kuwait was straining both the market and the organization. Fortunately for OPEC the destruction of Kuwaiti oilfields and sanctions against Iraq removed enough capacity long enough that the other members of the grouping had no difficulty achieving relatively stable markets until the late 1990s. Indeed, in 1996 prices rose sharply as OPEC capacity was inadequate, especially outside Saudi Arabia, even as some members, particularly Venezuela, expanded production aggressively, demonstrating that organizational cohesion is of secondary importance in a tight market characterized by low inventories and surplus capacity. The 1997 Asian economic crisis found OPEC facing a weak market, even as quota discipline reached new lows.

Growing Pie versus Appetite

While many will argue about the specific numbers, few would disagree that over the long run global demand will grow faster than non-OPEC supply.

However, while OPEC faces the pleasant task of dividing up an ever-growing pie, the members also have a ravenous appetite for revenues, which is unlikely to be sated more than briefly. Even the wealthiest OPEC members have ever-growing populations and social spending beyond what future oil revenues will cover, while many are in very poor financial straits. Indonesia, Iran, Nigeria, Qatar and Venezuela all have new governments that need to deliver improved economic performance to meet the expectations of their constituents. Outside of their oil sectors, there is not much potential for rapid improvement. This means that all these countries will seek higher oil revenues, which must come primarily from higher sales.

Table 8.1 Demand Growth for OPEC Oil:
Annual Change Under Different Scenarios
(mb/d, 1999–2004)

| | | Demand Growth | |
	mb/d	*Low*	*High*
Non-OPEC Supply		**1.5**	**2.0**
Low growth, no FSU	0.6	0.9	1.4
Low growth, plus FSU	0.9	0.6	1.1
High growth, no FSU	1.1	0.4	0.9
High growth, plus FSU	1.4	0.1	0.6

Source: Forecast by the author

Table 8.1 combines various non-OPEC supply scenarios with two demand scenarios for the coming five years. The low non-OPEC growth scenario (0.6 mb/d/yr) is slightly below the experience of the early 1990s, while the high non-OPEC growth (1.1 mb/d/yr) matches the experience of 1994–97, when cost cutting and new technologies resulted in greater growth in competitive supply. The growth in FSU capacity can be achieved if the pipeline from Kazakhstan is completed as planned and relatively minor incremental amounts from Russia and Azerbaijan are added. The two demand growth scenarios reflect the experience of the early 1990s (1.5 mb/d/yr) and the 1994–97 period (2.0 mb/d/yr).

The high non-OPEC plus FSU supply is most likely for the next five years, but so is the high demand scenario, meaning that OPEC should face incremental demand of the order of 0.6–0.8 mb/d/yr. This is healthy, but not robust, as the non-core OPEC members with active foreign upstream investment (Algeria, Iran, Libya, Nigeria, Qatar) will be adding between 0.3 mb/d/yr and 0.5 mb/d/yr. Assuming Iraq will want 0.2 mb/d/yr in new exports, there is not very much potential sales increase left for Kuwait, Saudi Arabia and Venezuela.

Combined with short-term and seasonal fluctuations caused by recessions, Caspian pipeline start-up or other factors, temporary production cutbacks will occasionally be needed, meaning that OPEC's internal cohesion remains important. While that cohesion is strong now, growing revenue needs and domestic political pressures on many members will make it difficult to maintain. In addition, the moderate growth in demand for OPEC oil means that it might be necessary for the smaller members to produce below capacity, which is problematic because of their increasing reliance on foreign oil companies for upstream investment. They will have to choose between forcing their foreign operators to delay investment or under-produce and thereby violate some production agreements, and having their national oil companies accept a disproportionate share of the production cutbacks. None of these options is very attractive, which will further complicate OPEC's job.

OPEC's New Management Strategy

OPEC's biggest challenges, notably competition and periodic market weakness, are beyond its control, but the organization has adopted a new market-management strategy which will alleviate some of its problems. When oil prices move outside the $22–28 range for more than 20 consecutive trading days, members will alter production by 500 tb/d (pro-rated).

This represents an improvement over current behavior, because by defining in advance both the point of action and the amount of production change, lengthy delay is avoided. This will minimize some of the volatility that results from both the market uncertainty about collective action and the higher inventories, which tend to occur when OPEC delays a cut in over-production while its members debate the situation or, even worse, try to schedule a debate. Better production data would also reduce uncertainty somewhat. Since volatility is self-reinforcing, the effect of less instability might be multiplied.

However, lack of enforcement powers leaves the other cause of volatility, cheating, unresolved. The recent experience of a major price collapse should in theory encourage voluntarism by the members but is no substitute for good, old-fashioned disciplinary action. After all, 1998 demonstrated that the members had forgotten the lesson of the 1986 price war, and by February 2000, quota discipline was said to be down to 74 percent. The lack of effective production data even reduces the effectiveness of moral persuasion. Perhaps the current spirit of cooperation makes this an auspicious time to reinstitute outside audits.

Previous oil groupings had either contractual or judicial power, including the Texas National Guard, to enforce production agreements. For OPEC to have equivalent power will require a major transformation, which seems highly unlikely. Even so, it pays to recall that it has managed weak markets before, not always well but always well enough. Achieving consensus may be difficult, but the difference in oil prices resulting from OPEC cooperation represents a lot of revenue.

The Future

Future oil-price stability will be a function of how the market will behave if left unimpeded, and then what price stabilization measures can be expected and how effective they will be.

The Market

The primary reason that price volatility is so much higher now is because of changes in the nature of the market, including market transparency, but also the industry's return to equilibrium. Understanding these will help to provide some insight into future price volatility as well as what is required to offset it.

Return to Equilibrium

The current refinery capacity constraint has been widely remarked upon, but there does not seem to be much appreciation of the degree to which this is a new phenomenon. In the past 20 years nearly all sectors of the petroleum industry have suffered from large amounts of surplus capacity. Figure 8.18 shows regional refinery capacity utilization, which has increased enormously after the collapse in the early 1980s, partly as refineries were closed in the OECD, partly as demand growth returned following the 1986 price collapse. If the Asian economic collapse had not occurred, the major refining centers would all be above 90 percent for the past few years, which is the same as full utilization, allowing for maintenance and other disruptions. The same is generally true of the global tanker industry and, as Figure 8.15 showed, OPEC's ability to produce crude oil.

Normally, this behavior means that the industry is optimally efficient, with minimal waste. However, when demand surges for whatever reason, whether because of strong economic growth or cold weather etc., or if supply is disrupted, whether due to politics, policy changes, accidents or any other reason, then there is little surge capacity available. This is compounded by the industry's decision of the mid-1990s to maintain so-called "just-in-time" inventories, or the minimum needed to operate efficiently (see Figure 8.19).

Given volatility of demand and large uncertainties about future demand, having minimal inventories and surge capacity means that prices will tend to be volatile, which is essentially what has happened since 1995. Before 1995 prices tended to drop due to overproduction and then recover. After 1995 the market has been also prone to swings above the long-term mean.

Figure 8.18
Refinery Capacity Utilization

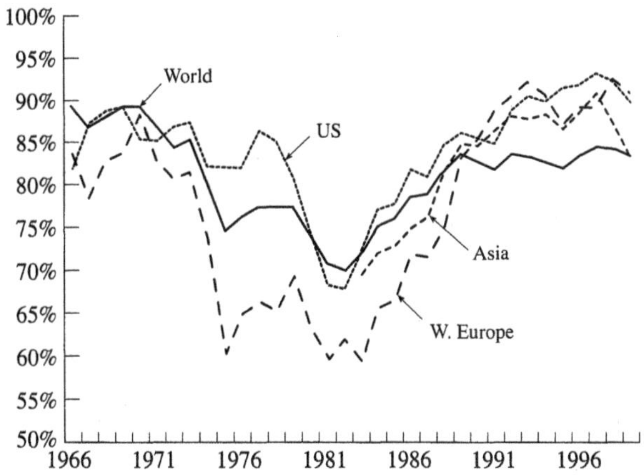

Source: BP Amoco, various issues

Figure 8.19
Usable Commercial Inventories

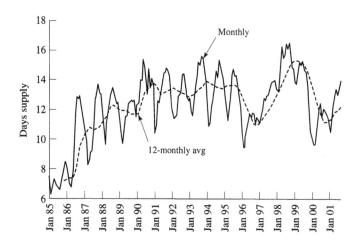

Source: *Oil Market Intelligence*, various issues

Market Transparency

One factor that is much overlooked is the decreasing physical transparency in the market due to the growing role of developing countries, where data are not collected in a timely fashion, if at all. For many years consumption has been growing faster in those countries than in the OECD, as Figure 8.20 shows. Yet the data in many of these countries tend not only to be unreliable and suffer from reporting lag, but often are only reported on an annual basis. Also, many of these countries still undertake short-term policy interventions in the market, attempting to influence consumption, imports and/or prices in ways that are not only poorly reported but can have a significant short-term impact on the market. For example, the changing of oil product price controls or altering of subsidies still occurs in a number of large consuming countries in the developing world.

Most importantly, a growing percentage of the world's inventories are apparently located in these regions (Figure 8.21), but there is no measurement of them. Market observers who report inventory levels, such as the IEA and OMI, are using a formula under which they have estimated consumption in those areas, a seasonality adjustment is applied, and that is multiplied by 55 days. The latter number is derived from an old study by Royal Dutch/Shell, which found that, at the time of the study, non-OECD

Figure 8.20
Relative Oil Consumption: OECD and LDCs

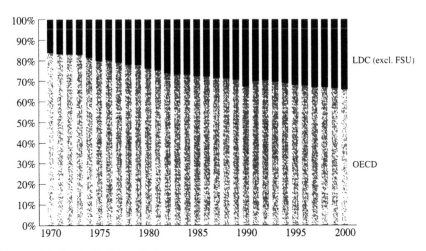

Source: USDOE, *Monthly Energy Review*

inventories held enough oil for 55 days of consumption. The number has never been updated, nor has there been any effort to estimate the dynamic behavior of the number: how much it varies and why. Yet, typically, changes in the estimated inventory levels in these areas are reported as actual inventory changes.

This goes a long way toward explaining the current disparity between high oil prices and OPEC's fears of overproducing and creating another price collapse. There is serious uncertainty about the trend in inventories over the next few months, since the production and consumption data suggest they should have been increasing much more recently than they have been (the "missing barrels" shown in Figure 8.11).

What can be done?

The primary challenge is that oil markets are volatile and becoming more so, and it is not clear if OPEC has the understanding, the ability or the power to stabilize markets, mainly because it is not clear if anyone could have the understanding, the ability or the power to stabilize markets. Coping with each of the different problems requires different responses, as discussed below.

Figure 8.21
Location of Commercial Inventories (Share of the Total)

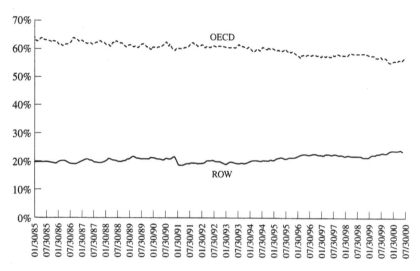

Source: *Oil Market Intelligence*, various issues

Market Volatility

There is no feasible method that will make economic growth more stable or reduce the volatility of weather, at least none not already in place for other reasons. Similarly, the private sector cannot be expected to hold expensive surplus capacity or high inventory levels in order to reduce price volatility, especially as these primarily reduce the occasional price spike and increase the likelihood of price troughs. Unless OPEC or the IEA wishes to pay to maintain 5–10 percent capacity throughout the industry (crude, refiners, tankers, inventories), fluctuations in demand and, to some degree, in supply must simply be accepted as part of the background conditions of the industry.

On the other hand, the poor transparency could be improved in a number of ways. OPEC nations could allow outside audits of their production not only to improve the accuracy of the data, but also to raise the confidence felt by the oil market in the accuracy of that data. Other countries could be urged to improve reporting of demand and inventory data, and possibly outside nations (IEA or OPEC) could provide funding and other assistance in procuring this data.

Internal Management

Price (and therefore revenue) volatility is a challenge to an organization such as OPEC. The most profitable behavior for producers in any cartelized market is to cheat. Combined with the various market uncertainties, OPEC therefore has a difficult job that it is ill-suited to do. The Texas Railroad Commission had the power to compel not only accurate information from its producers, but compliance with production quotas too. The Seven Sisters were partners in a variety of interlocking ventures, so that they had full access to the bulk of each other's data. The contracts were also designed to prevent overlifting.

OPEC has none of the advantages of the other groups that have controlled the market and instead must rely on voluntarism. Forcing the members to accept outside auditors and having production decisions turned over to it would improve its ability to manage the market. The latter in particular is unlikely, but the recent adoption of a trading band is a step toward improving that situation.

Yet, given the complexity and enormity of the volatility of the oil market, it seems clear that the best OPEC, or any other body, can do is to attempt to reduce market uncertainties and price fluctuations, not eliminate them. However, even that could yield substantial economic benefits.

Conclusions

Stability in the oil industry is likely to be a historical artifact or sought-after goal, but is unlikely to be achieved in the future, except in a relative sense. Some things can be done to improve it, including better data collection and reporting. However, it is not likely that OPEC nations will yield control over their oil production, which means that their quota discipline will always be imperfect. In addition, quotas themselves will always include some degree of error, given the natural volatility of commodities.

Thus, producers and consumers should accept some volatility as part of the natural order in the oil market and realize that a high level of stability requires changes that are very expensive and/or politically unacceptable, while trying to reduce the uncertainty as much as possible.

Concluding Observations

Robert Mabro

The eight chapters included in this book, together with the introduction, address a question of vital importance to both oil-exporting countries depending on petroleum revenues for their economic well-being and oil companies which explore, develop, refine and market a commodity so essential to our modern lifestyle. The question simply is whether oil has a future as a source of energy.

The issue is examined under three aspects. Since oil is an exhaustible natural resource, the depletion of the reserves sets an end-date to the production and use of oil. The first aspect, therefore, relates to geology.

However, oil could be displaced by substitutes before the end of its geophysical life. This second and more important aspect is covered in four chapters that assess the prospects of alternative fuels to oil.

The third part focuses on concerns other than those caused by the potential threat of alternative energies. Environmental policies, for example, may impact on the demand for oil, not only by encouraging the development of renewables, but also by introducing petroleum-saving technologies in the transport sector. The instability of oil markets is a different and more subtle source of concern. It may aggravate fears about the security of supplies and strengthen the hand of those who argue that it is too risky to depend heavily on oil. This could increase the impetus in the search for alternatives and for greater efficiency in the use of energy.

The third part includes a chapter on new technologies in the oil industry. These tend to reduce the costs of exploration, development and production. However, their effects are ambiguous. On the one hand, the reduction in costs will threaten the economic viability of substitutes and improve the future prospects of oil. On the other hand, the new technologies help to increase non-OPEC production and to that extent pose a threat, not to oil in general but to the stability and welfare of the exporting countries.

Let us now consider each chapter in turn.

On the geological prospects, in the first chapter in this book, Mr. Jean Laherrère warns us that data on oil reserves are so poor as to make forecasting, a difficult exercise in any case, even more unreliable. Data reporting, be it by countries or companies, is influenced by politics. Detailed critical analysis that the author applies carefully leads him to forecast that a production peak of conventional oil will be reached around 2010. After that, the main sources of supply will come from the Gulf region, the Orinoco and Athabasca belts and from liquids derived from wet natural gas. However, given the quality of the data, about which the author rightly complains, we have to suspend judgment. We can only wait for the day when the relevant authorities will be persuaded to provide statistics free from political interference to assess with some confidence the geophysical future of oil.

On alternatives to oil, a chapter by Dr. David Hart provides a general assessment that "seeks to understand some of the drivers of change within energy markets, from an economic, environmental, political and technical perspective."

It is true that a complete set of forces that want to drive the energy world away from oil is already at work and may gain momentum. Yet, the obstacles in their way, which will result in very long lead-in times, are also formidable. Dr. Hart sees a future in which the twin carriers of electricity and hydrogen will "link together fossil and non-fossil energy structures." They may, however, as he puts it, help the move from one to the other, which I assume to imply a move away from the fossil sources of energy.

Dr. Timothy Lipman is more explicitly optimistic about the potential for renewables. Their promise "has now become a reality." Solar photovoltaics and wind energy are benefiting from cost reductions and their market is expanding. However, the potential, to be realized, requires much more investment than is currently being undertaken. He advises current energy producers to use their resources to capture new renewable energy sources.

Governments may have to provide some encouragement. If Lipman is right that the promise of renewables is no longer a hope for the distant future, then entrepreneurs will sooner or later recognize the opportunities and try to seize them. The author does not tell us why there is so much apparent hesitation on their part.

Natural gas has been considered for a long time now as the heir of oil. The transition from the oil age to another energy age will involve gas. Dr. Marian Radetzki's view that the future of natural gas is quite bright suggests his agreement with the conventional wisdom. He argues the case by referring to the historical growth in gas consumption, the availability of large reserves and recent forecasts about the future demand for gas, which, although predicting robust growth, are considered to be too conservative by the author. Finally, Dr. Radetzki sees a possible breakthrough in the gas-to-liquid markets leading to large-scale displacement of middle distillates.

Given that, some time in the future, a transition from fossil fuels (not only oil, but coal and gas) to non-fossil alternatives is inevitable, the question of "what challenges such a transition involves" does arise. This important issue is addressed by Dr. Cutler Cleveland. He argues against the view that energy use and economic well-being will be decoupled in the future, making the transition relatively easy. His point is that the quality of solar energy is inferior to that of fossil fuels: less energy density, lower energy return on investment, high energy input in the delivery system relative to the amounts delivered. These have economic implications. The decoupling between energy and economic growth may not occur.

The future of oil is threatened by environmental concerns. The nature of these concerns and their impact on the future energy mix are comprehensively dealt with by Mr. Seth Dunn in his chapter. The importance of good husbandry, of ensuring that we do not pollute and damage our planet more than our activities already have is well recognized. However, what is lacking is the political will to undertake the necessary remedial action. This is the crux of the matter. We need to hear more about what can and should be done to persuade those who can make sacrifices to accept them and those who should lead international cooperation to move to the front.

An optimistic note about the future of oil is sounded in Dr. Kelkar's chapter, which discusses the new technological developments that enable the industry to extract more hydrocarbons at lower costs in hostile environments. The author recognizes that the future challenges are daunting but believes that the oil industry can face them successfully. However, the problem is that the threat to oil does not come from within (geology for example) but from outside: environment, politics and technical progress in other sectors. To be able to produce more oil cheaply may retard substitution but does not address the other challenges.

Finally, Mr. Michael Lynch analyses the issue of oil market stability. His essay is largely historical. Volatility has always been there throughout the various episodes of oil history. He concludes that one needs to accept it as inevitable but to adopt policies that reduce somewhat certain uncertainties. The implications of this instability for the future of oil are not mentioned.

Given that the future is uncertain, the only sensible conclusion is that oil-exporting countries would be well advised to embark without delay on policies that improve the prospects of economic development with reduced dependence on oil. Failure to face this challenge will inevitably mean that their oil era, even if it lasts for a very long time, will be followed by an era of growing poverty.

CONTRIBUTORS

CUTLER J. CLEVELAND is Director and Professor of the Center for Energy and Environmental Studies at Boston University. He holds a B.S. in Ecology and Systematics from Cornell University, an M.S. in Marine Science from Louisiana State University and a Ph.D. in Geography from the University of Illinois at Urbana-Champaign. He is a member of the Scientific Planning Committee for the Human Dimensions of Global Environmental Change/Industrial Transformation Program.

Dr. Cleveland has been a consultant to numerous private and public organizations, including the Asian Development Bank, the Technical Research Center of Finland, the US Department of Energy and the US Environmental Protection Agency. Dr. Cleveland's research focuses on the ecological-economic analysis of how energy and materials are used to meet human needs. His research employs the use of econometric and systems dynamics models of oil supply, natural resource scarcity and the relation between the use of energy and natural resources and economic systems.

SETH DUNN is Research Associate with the Worldwatch Institute, an environmental policy think tank based in Washington, DC. He is a graduate of Yale University, where he studied history and environmental studies. Mr. Dunn has authored or co-authored four editions of the Institute's *State of the World* annual and three Worldwatch Papers, including most recently *Hydrogen Futures: Toward a Sustainable Energy System* (August 2001). He has contributed over 30 book chapters and articles to outside publications. His work has been cited in numerous outlets, including *The Economist, Nature, New Scientist*, the *New York Times* and the *Washington Post*.

DAVID HART is Head of Fuel Cells and Hydrogen Research at the Imperial College Centre for Energy Policy and Technology (ICCEPT) in London, specializing in fuel cell use in both decentralized generation and in transportation applications. The main thrusts of his work concern the economics, policy and environmental implications of fuel cell and hydrogen energy use in a wide variety of different contexts, and the interfaces both with renewable energy and with conventional fossil-fuel sources.

His work at Imperial College follows on from a postgraduate degree in Environmental Technology and Energy Policy, specializing in fuel cells and microturbines in small-scale power generation, and previous work in Germany and Japan as a control systems engineer. He was a co-author of the recent Financial Times Energy management report *Decentralized Electricity*, and has also authored and co-authored reports on fuel cells and hydrogen power. He has consulted for a number of major organizations and is a Director of E4tech, a specialist Energy Environment consultancy.

DANIEL M. KAMMEN received his doctorate in physics from Harvard and was the Wezmann & Bantrell Postdoctoral Fellow at the California Institute of Technology in the Divisions of Engineering, Biology, and the Humanities. At Caltech and then as a Lecturer in Physics in the Kennedy School of Government at Harvard University, Dr. Kammen developed a number of projects focused on renewable energy technologies and environmental resource management. At Harvard he also worked on risk analysis as applied to global warming and methodological studies of forecasting and hazard assessment. Dr. Kammen received the 21st Century Earth Award 1993, recognizing contributions to rural development and environmental conservation, from the Global Industrial and Policy Research Institute and *Nihon Keizai Shimbun* in Japan.

As an Assistant Professor in the Woodrow Wilson School of Public and International Affairs at Princeton University, Dr. Kammen played a key role in developing the interdisciplinary Science, Technology and Environmental Policy (STEP) Program, serving both as Co-Chairman and Chairman of STEP. In 1998 he joined the Energy and Resources Group (ERG) at Berkeley as an Associate Professor of Energy and Society and is also the Founding Director of a new Renewable and Appropriate Energy Laboratory (RAEL). Dr. Kammen is Associate Professor of Nuclear Engineering and faculty member in the Health, Environment and Development Program, and the Committee on African Studies. He has authored over 100 articles, 4 books, 5 technical reports, over 40 abstracts and has reviewed a number of conference publications and popular articles.

MOHAN KELKAR is Professor of Petroleum Engineering at the University of Tulsa. He has a B.S. degree from the University of Bombay, an M.S. degree and a Ph.D. in chemical engineering from the University of Pittsburgh, and a J.D. degree from the University of Tulsa. He has authored or co-authored over 40 academic articles and is currently busy with a book on *Application of Geostatistics for Reservoir Characterization*, to be published by the Society of Petroleum Engineers. He is a member of the American Institute of Chemical Engineers and the Oklahoma Bar Association, and he has served on the SPE Editorial Review Committee. Currently, he participates in several research projects regarding reservoir characterization, which are jointly funded by the US Department of Energy and oil companies.

JEAN LAHERRÈRE graduated from the École Polytechnique and the École Nationale du Petrole in Paris. He joined the Compagnie Française des Petroles (now TOTAL) in 1955 and participated in the exploration of Sahara and Australia. He was in charge of exploration in Canada for TOTAL in Calgary. After 15 years overseas, he went to the TOTAL headquarters in Paris where he was successively in charge of new ventures negotiation, technical services and research, basin exploration departments and finally Deputy Exploration Manager. He was a member of the Safety Panel of the Ocean Drilling Program (JOIDES). He was President of the Exploration Commission of the Technical Committee of the French oil industry union, managing several manuals on exploration. He was Director of the Compagnie Générale de Géophysique, Petrosystems and various TOTAL subsidiaries. He retired in 1991.

Jean Laherrère is now writing articles and giving lectures. He has written several reports with Petroconsultants and *Petroleum Economist* on the world's oil and gas potential and future production. He was a member of the Society of Petroleum Engineers/World Petroleum Congress ad hoc Committee on joint definitions of petroleum reserves and also a member of the task force on Perspectives Énergie 2010-2020 for the Commissariat Général du Plan. He will chair the panel on hydrates at the World Petroleum Congress, 2002.

TIMOTHY E. LIPMAN is currently Post-doctoral Research Fellow with the Energy and Resources Group and the Renewable and Appropriate Energy Laboratory (RAEL) at the University of California at Berkeley. He is leading a research project examining the economics and environmental impacts of using fuel cells as distributed power-generating resources. Dr. Lipman completed a Ph.D. degree in Environmental Policy Analysis with the Graduate Group in Ecology at the University of California at Davis. He then served as Associate Director of the Fuel Cell Vehicle Center at the Institute of Transportation Studies (ITS), UC Davis. After completing an M.S. degree in technology, Dr. Lipman worked as a researcher at ITS-Davis on a CALSTART neighborhood electric vehicle project and on a California Air Resources Board electric vehicle cost and performance study, as well as various smaller projects. He has authored or co-authored several journal articles and ITS-Davis research reports.

A graduate of Stanford University, Dr. Lipman has previously worked as an environmental analyst and technical writer with the NASA-Ames Research Center in Mountain View, California, and as a policy intern at the Bank Information Center in Washington, DC. He has received several fellowships and awards, including the 1999 Council of University Transportation Center Charley Wootan best dissertation award, a 1998 IGERT teaching fellowship, a 1997 University of California Transportation Center Dissertation Grant, a 1996 Eno Transportation Foundation Fellowship, a 1995 University of California Transportation Center Dissertation Grant, and a 1994 Chevron Foundation Fellowship.

MICHAEL C. LYNCH is Chief Energy Economist at DRI-WEFA, Inc., a leading economic consulting firm and a research affiliate at the Massachusetts Institute of Technology's Center for International Studies. He has combined Bachelor of Science and Master of Science degrees in Political Science from MIT and has undertaken a variety of studies related to international energy matters, including forecasting of the world oil market, energy and security and corporate strategy in the energy industries, as well as an analysis of oil and gas supply.

Mr. Lynch is a past President of the United States Association for Energy Economics, and was the Program Chairman of their 1996 North American Conference. He was appointed council member of the International Association for Energy Economics. His publications have appeared in Spanish, Arabic, Italian, Russian and Japanese, as well as English, and he serves on the editorial board of the journal *Energy Policy* as well as on Pennwell Publishing's Petroleum Technical Review Board.

ROBERT MABRO CBE, Fellow of St. Antony's College, Oxford, is the Director of the Oxford Institute for Energy Studies. After obtaining a degree in civil engineering from Alexandria University, he worked as a civil engineer in

Egypt. Later he studied philosophy in France and obtained an M.Sc in Economics with distinction from London University. Mr. Mabro began his academic career at the School of Oriental and African Studies (SOAS) at London University, specializing in Middle East economic developments. In 1969 he joined Oxford University as Senior Research Officer in Middle East economics. Mr. Mabro's interest in oil began to develop from 1972, and his first publication in this field was a monograph co-authored with the late Elizabeth Monroe entitled *Oil Producers and Consumers: Conflict or Cooperation* (New York, 1974). His list of publications includes 13 books and monographs as well as numerous articles and papers. His most recent book, written with Dr. Paul Horsnell, is *Oil Markets and Prices: The Brent Market and the Formation of World Oil Prices* (Oxford University Press, 1993).

Mr. Mabro co-founded the Oxford Energy Policy Club at St. Antony's College, and also served as founder-director of the annual Oxford Energy Seminar at St. Catherine's College. The next step was the establishment of the Oxford Institute for Energy Studies, an educational charity devoted entirely to research on the economics, politics and international relations of energy. He was awarded the International Association for Energy Economics award in 1990 for his outstanding contributions. In December 1995 he was awarded a CBE by Her Majesty the Queen, in the New Year Honors List. The President of Mexico awarded him the medal of the Mexican Order of Aguila Azteca in 1997 and the President of Venezuela awarded him the medal of Francisco Miranda in 2000.

MARIAN RADETZKI is Professor of Economics at Luleå University of Technology in Sweden and Senior Research Fellow at SNS, a Swedish think tank. His research focuses on energy and mineral economics and more recently also on environmental economics. Radetzki has published widely in the field. He has held positions as visiting Professor of Mineral Economics at the Colorado School of Mines and as Chief Economist at CIPEC, the Intergovernmental Council of Copper Exporting Countries.

Dr. Radetzki has consulted for several international organizations, including the World Bank, as well as for private corporations. His book *The Green Myth – Economic Growth and the Quality of the Environment*, recently published by Multiscience Publishing Co. in the United Kingdom, is scheduled for publication in Arabic by ECSSR.

Introduction

1   Robert Mabro, "The World's Oil Supply 1930–2050," A Review Article, *The Journal of Energy Literature*, Vol. II, No. 1 (1996): 32.

Chapter 1

1   J.H. Laherrère, "Assessing the Oil and Gas Future Production and the End of Cheap Oil?" *CSEG*, Calgary April 6, 1999 (http://dieoff.com/page179.htm).
2   Walter Youngquist, "Shale Oil: The Elusive Energy," *Hubbert Center Newsletter* April 1998 (http://hubbert.mines.edu).
3   M. Hovland, "Are There Commercial Deposits of Methane Hydrates in Ocean Sediments?" JNOC. "Methane Hydrates: Resources in the Near Future?" JNOC-TRC, October 20–22, 1998 (http://www.simmonsco-intl.com/web/downloads/spe.pdf); J.H. Laherrère, "Oceanic Hydrates: More Questions Than Answers," *Energy Exploration and Exploitation*, Special Issue on Hydrates, 2000.
4   G.D. Ginsburg, "Challenging the Presence of Natural Gas Hydrate in the Messoyakha Pool," *AAPG* vol. 77, no. 9 (1993).
5   *Petroleum Economist* (May 2000): 37.

Chapter 2

1   A. Grübler, N. Nakicenovic and D. Victor, "Dynamics of Energy Technologies and Global Change," *Energy Policy* vol. 27, no. 5 (1999).
2   F. Pearce, "Back to the Days of Deadly Smogs," *Nature* no. 1850 (1992).
3   W. Blackburn, "Strategies for Commercializing Fuel Cell Vehicles in California," *Fuel Cell Seminar 2000*, Portland, Oregon, 2000.
4   J.T. Houghton, L.G.M. Filho, B.A. Callender, N. Harris, A. Kattenberg and K. Maskell, IPCC Second Assessment Report, *Climate Change*

1995: The Science of Climate Change (Bracknell, UK: Meteorological Office, 1995).

5   Albritton, D.L. et al., *Summary for Policymakers: A Report of Working Group I of the Intergovernmental Panel on Climate Change*, International Panel on Climate Change (Port Chester, NY: Cambridge University Press, 2001).

6   European Commission, *Energy for the Future: Renewable Sources of Energy* (Brussels, Belgium: European Commission, 1997).

7   D. Papathanasiou and D. Anderson, *Uncertainties in Responding to Climate Change: On the Economic Value of Technology Policies for Reducing Costs and Creating Options* (London: Imperial College Centre for Energy Policy and Technology, 2000).

8   J.M. Ogden, "Developing an Infrastructure for Hydrogen Vehicles: A Southern California Case Study," *International Journal of Hydrogen Energy*, vol. 24, no. 8 (1999); C. Thomas, B. James, F. Lomax and I. Kuhn, "Integrated Analysis of Hydrogen Passenger Vehicle Transportation Pathways," *DE–AC36–83CH10093* (National Renewable Energy Laboratory, 1998).

9   W. Heuer, "The TES Project: A Joint Initiative for an Additional Fuel Infrastructure," *Journal of Power Sources* vol. 86 (2000).

10   E.F. Schumacher, *Small is Beautiful* (London: Vintage, 1993).

11   J. Crosse, "Almost the Real Thing," *FT Automotive World*, May 1999, 40–43.

12   N. Nakicenovic, A. Grübler and A. McDonald, *Global Energy Perspectives* (Cambridge, UK: Cambridge University Press, 1998).

13   See IIASA website at http://www.iiasa.ac.at/cgi-bin/ecs/book_dyn/bookcnt.py.

14   S.A. Moss, "Potential Carbon Emission Savings from Energy Efficiency in Commercial Buildings," *IP3/96* (Watford: Buildings Research Establishment, 1996).

15   W. Patterson, *Transforming Electricity* (London: Earthscan, 1999).

16   Council of the European Union, Council Directive 99/32/EC 1999.

17   P. Tempest, *World Petroleum at the Crossroads* (London: PTA Greenwich, 1999).

18   See Chapter 3.

19   D. Hart and A. Bauen, *Fuel Cells: Clean Power, Clean Transport, Clean Future* (London: Financial Times Energy Publishing, 1997); K.

Kordesch and G. Simader, *Fuel Cells and Their Applications* (Weinheim, Germany: VCH Verlaggesellschaft, 1996); J. Larminie and A. Dicks, *Fuel Cell Systems Explained* (Chichester: Wiley, 2000).

20 A. Bauen and D. Hart, "Assessment of the Environmental Benefits of Transport and Stationary Fuel Cells," *Journal of Power Sources* vol. 86, no. 1–2 (2000); B. Höhlein, S. von Andrian, T. Grube and R. Menzer, "Critical Assessment of Power Trains with Fuel Cell Systems and Different Fuels," *Journal of Power Sources* vol. 86, no. 1–2 (2000); C. Thomas, B. James, F. Lomax and I. Kuhn, "Integrated Analysis of Hydrogen Passenger Vehicle Transportation Pathways," *DE–AC36–83CH10093* (National Renewable Energy Laboratory, 1998).

21 A. Grübler, N. Nakicenovic and D. Victor, "Dynamics of Energy Technologies and Global Change," *Energy Policy* vol. 27, no. 5 (1999).

22 G.D. Rambach, "An Examination of Isolated, Stationary, Hydrogen Power Systems Supplied by Renewables," *9th Canadian Hydrogen Conference*, Vancouver, Canada, 1999.

23 C. Gregoire Padró and V. Putsche, *Survey of the Economics of Hydrogen Technologies, NREL/TP–570–27079* (National Renewable Energy Laboratory, 1999).

24 Ibid.

25 D. Hart, M.A. Leach, R. Fouquet, P.J. Pearson and A. Bauen, "Methanol Infrastructure: Will it Affect the Introduction of SPFC Vehicles?" *Journal of Power Sources* vol. 86, no. 1–2 (2000).

26 L. Hartley, *The Go-between* (Harmondsworth, UK: Penguin, 1999).

27 J.H. Wang, W.L. Chiang and J.P.H Shu, "The Prospects: Fuel Cell Motorcycle in Taiwan," *Journal of Power Sources* vol. 86, no. 1–2 (2000).

28 Fagan, M. "Sheikh Yamani Predicts Price Crash as Age of Oil Ends," *The Sunday Telegraph*, London, June 25, 2000.

Chapter 3

1 President's Committee of Advisors on Science and Technology (PCAST), *Federal Energy Research and Development for the Challenges of the Twenty-First Century,* Report of the PCAST Energy

Research and Development Panel (Washington, DC: November 1997); A. Kinzig and D. Kammen, "National Trajectories of Carbon Emissions: Analysis of Proposals to Foster the Transition to Low-Carbon Economies," *Global Environmental Change* vol. 8, no. 3 (1998); Shell Petroleum, *Global Scenarios: 1998–2020,* Summary Brochure, 2000.

2   Amory Lovins, *Soft Energy Paths: Toward a Durable Peace* (Cambridge, MA: Ballinger Press, 1977); Joan Ogden and Robert Williams, *Solar Hydrogen: Moving Beyond Fossil Fuels* (Washington, DC: World Resources Institute, 1989); Christopher Flavin and Nicholas Lenssen, *Power Surge: Guide to the Coming Energy Revolution* (New York, NY: W.W. Norton and Co., 1994); Thomas Johansson, Henry Kelly, Amulya Reddy and Robert Williams (eds) *Renewable Energy: Sources for Fuels and Electricity* (Washington, DC: Island Press, 1993).

3   Jamie Chapman, Steven Wiese, Edgar DeMeo and Adam Serchuk, *Expanding Wind Power: Can Americans Afford It?*, Renewable Energy Policy Project Report No. 6, November 1998.

4   American Wind Energy Association, "Wind Energy Fact Sheet: Comparative Costs of Wind and Other Fuels" (URL http://www.awea.org) 2000.

5   US Department of Energy, *Renewable Energy Technology Characterizations* (Washington, DC: USDOE Office of Utility Technologies and EPRI, 1997).

6   Ibid.

7   Chihiro Watanabe, "Industrial Dynamism and the Creation of a 'Virtuous Cycle' between R&D, Market Growth and Price Reduction: The Case of Photovoltaic Power Generation (PV) Development in Japan," in Clas-Otto Wene, A. Voss and T. Fried (eds) *Proceedings of the IEA International Workshop on Experience Curves for Policy Making: The Case of Energy Technologies, Stuttgart, Germany, May 10–11, 1999* (Stuttgart: 2000).

8   US Department of Energy, op. cit. (Note 5).

9   Ibid.

10 B.G. Swezey and Yih-huei Wan, *The True Cost of Renewables: An Analytic Response to the Coal Industry's Attack on Renewable Energy* (Golden, CO: US Department of Energy, 1996).

11 Jamie Chapman, Steven Wiese, Edgar DeMeo and Adam Serchuk, *Expanding Wind Power: Can Americans Afford It?*, Renewable Energy Policy Project Report No. 6, November 1998.

12 Richard Ottinger, David Wooley, David Hodas, Nicholas Robinson and Susan Babb, *Environmental Costs of Electricity* (New York, NY: Oceana Publications, Inc., 1991).

13 Paul Maycock, "The World PV Market 2000: Shifting from Subsidy to 'Fully Economic'?" *Renewable Energy World* vol. 3, no. 4 (2000).

14 BTM Consult APS, "Wind Energy Statistics" (http://www.btm.dk/Statistics.htm) 2000.

15 Shell Petroleum, op. cit. (Note 1).

16 Reuters News Service, "Fuel Cells and New Energies Come of Age Amid Fuel Crisis" (http://www.reuters.com) September 11, 2000.

17 Nuveen Investments, *Nuveen Defined Portfolios Product Guide*, company literature, June 2000.

18 Richard Duke and Daniel Kammen, "The Economics of Energy Market Transformation Initiatives," *The Energy Journal* vol. 20 (1999); Richard Duke, S. Graham, M. Hankins, A. Jacobson, D. Kammen, D. Khisa, D. Kithokoi, F. Ochieng, B. Osawa, S. Pulver and E. Walther, *Field Performance Evaluation of Amorphous Silicon (a-Si) Photovoltaic Systems in Kenya: Methods and Measurements in Support of a Sustainable Commercial Solar Energy Industry* (Washington, DC: The World Bank, 2000); Daniel Kammen, "Bringing Power to the People: Promoting Appropriate Energy Technologies in the Developing World," *Environment* vol. 41, no. 5 (1999).

19 Intergovernmental Panel on Climate Change Working Groups II and III, *Methodological and Technological Issues in Technology Transfer* (Cambridge, UK: Cambridge University Press, 2000).

20 Robin Cowan and David Kline, *The Implications of Potential "Lock-In" in Markets for Renewable Energy*, International Symposium on Energy and Environmental Management, Newport Beach, California, 1996.

21 Jorge Islas, "Getting Round the Lock-In in Electricity Generating Systems: The Example of the Gas Turbine," *Research Policy* vol. 26 (1997).

22 W. Brian Arthur, "Competing Technologies, Increasing Returns, and Lock-In by Historical Events," *The Economic Journal* vol. 99 (March 1989).

23 Joan Ogden, Eric Larson and Mark Delucchi, *A Technical and Economic Assessment of Renewable Transportation Fuels and Technologies* (Washington, DC: Office of Technology Assessment, May 1994).

24 Timothy Lipman, *Zero-Emission Vehicle Scenario Cost Analysis Using a Fuzzy Set-Based Framework* (Davis, CA: Institute of Transportation Studies, December 1999).

25 Willett Kempton and Steven Letendre, "Electric Vehicles as a New Power Source for Electric Utilities," *Transportation Research – D* vol. 2, no. 3 (1997).

26 Timothy Lipman, *Zero-Emission Vehicle Scenario Cost Analysis Using a Fuzzy Set-Based Framework* (Davis, CA: Institute of Transportation Studies, December 1999).

27 Paul Baer, John Harte, B. Haya, Antonio Herzog, John Holdren, Nathan Hultman, Daniel Kammen, Richard Norgaard and L. Raymond, "Equity and Greenhouse Gas Responsibility," *Science* vol. 289 (2000).

28 *Federal Register*, "National Ambient Air Quality Standards for Particulate Matter: Proposed Decision," vol. 61 (1996).

29 T. Hughes, *Networks of Power: Electrification in Western Society, 1870–1930* (Baltimore, MD: Johns Hopkins University Press, 1983); Richard Margolis and Daniel Kammen, "Underinvestment: The Energy Technology and R&D Policy Challenge," *Science* vol. 285 (1999); R. Hirsh, *Power Loss: The Origins of Deregulation and Restructuring in the American Utility System* (Cambridge, MA: MIT Press, 2000).

30 Hirsh, op. cit. (Note 29).

31 Margolis and Kammen, op. cit. (Note 29).

32 M. Spence, "The Learning Curve and Competition," *Bell Journal of Economics* vol. 12 (1981).

33 Duke and Kammen, op. cit. (Note 18).

34 President's Committee of Advisors on Science and Technology (PCAST), op. cit. (Note 1).

Chapter 4

Acknowledgment: The author gratefully acknowledges valuable comments from participants to the conference at which the paper was first presented, and from two anonymous referees.

1 See International Energy Agency, *Natural Gas Information* (Paris: OECD, 1998a).
2 See International Energy Agency, *Oil, Gas and Coal Supply Outlook* (Paris: OECD, 1995).
3 IEA Monthly Oil Report, June 1998.
4 *World Gas Intelligence*, vol. 7, no. 20, October 11, 1996.

Chapter 5

Acknowledgment: The author gratefully acknowledges the comments from many of the participants at the Sixth Annual Energy Conference "The Future of Oil as an Energy Source," sponsored by the Emirates Center for Strategic Studies and Research, October 7–8, 2000 in Abu Dhabi, United Arab Emirates.

1 J. Gever, R. Kaufmann, D. Skole and C. Vorosmarty, *Beyond Oil: The Threat to Food and Fuel in the Coming Decades* (Cambridge, MA: Ballinger, 1986). See also C.A.S. Hall, C.J. Cleveland and R.K. Kaufmann, *Energy and Resource Quality: The Ecology of the Economic Process* (New York, NY: Wiley Interscience, 1986).
2 C.A.S. Hall, J.A. Stanford and F. Richard Hauer, "The Distribution and Abundance of Organisms as a Consequence of Energy Balances along Multiple Environmental Gradients," *OIKOS* vol. 65 (1992): 377–90.
3 H.T. Odum, "Trophic Structure and Productivity of Silver Springs, Florida," *Ecological Monographs* vol. 27 (1957): 55–112.
4 V. Smil, *General Energetics* (New York, NY: Wiley-Interscience, 1991).
5 Hall et al., op. cit. (Note 1).
6 C.J. Cleveland, R. Kaufmann and D.I. Stern, "Aggregation and the Role of Energy in the Economy," *Ecological Economics* vol. 32 (2000): 301–17.

7   F.G. Adams and P. Miovic, "On Relative Fuel Efficiency and the Output Elasticity of Energy Consumption in Western Europe," *Journal of Industrial Economics* vol. 17 (1968): 41–56.

8   E.R. Berndt, "Aggregate Energy, Efficiency, and Productivity Measurement," *Annual Review of Energy* vol. 3 (1978): 225–73 and 242.

9   R.K. Kaufmann, "The Relation Between Marginal Product and Price in US Energy Markets," *Energy Economics* vol. 16 (1994): 145–58.

10   Adams and Miovic, op. cit. (Note 7).

11   C.J. Cleveland, R. Costanza, C.A.S. Hall and R. Kaufmann, "Energy and the US Economy: A Biophysical Perspective," *Science* vol. 255 (1984).

12   D.I. Stern, "Energy Use and Economic Growth in the USA: A Multivariate Approach," *Energy Economics* vol. 15 (1993). See also R.K. Kaufmann, "A Biophysical Analysis of the Energy/Real GDP Ratio: Implications for Substitution and Technical Change," *Ecological Economics* vol. 6 (1992): 35–56.

13   Stern, op. cit. (Note 12) and Cleveland et al., op. cit. (Note 6).

14   Kaufmann, op. cit. (Note 12).

15   N. Georgescu-Roegen, "Energetic Dogma, Energetic Economics and Viable Technology," in J.R. Moroney (ed.) *Advances in the Economics of Energy and Resources* (Greenwich, CT: JAI Press, 1982), 30.

16   W.F. Cottrell, *Energy and Society* (New York, NY: McGraw-Hill, 1995); H.T. Odum, *Environment, Power and Society* (New York, NY: Wiley-Interscience, 1971); Cleveland et al., op. cit. (Note 11); C.J. Cleveland, "Energy Quality and Energy Surplus in the Extraction of Fossil Fuels in the US," *Ecological Economics* vol. 6 (1992).

Chapter 6

1   Jesse H. Ausubel, "Where is Energy Going?" *The Industrial Physicist* vol. 6, no. 1 (February 2000): 16–20.

2   Figure 6.1 is from Mac Post, "Global Carbon Cycle (1992–1997)," Carbon Dioxide Information Analysis Center (CDIAC), Oak Ridge National Laboratory (ORNL), viewed on September 7, 2000 (cdiac.esd.ornl.gov/pns/graphics/globcarb.gif).

3   Figure 6.2 is from G. Marland, T.A. Boden and R.J. Andres, "Global, Regional, and National CO_2 Estimates from Fossil Fuel Burning,

Cement Production, and Gas Flaring: 1751–1997" (revised August 2000), CDIAC, ORNL, Oak Ridge, TN, 22 August 2000, and in BP Amoco, *BP Amoco Statistical Review of World Energy 2000* (London: Group Media & Publications, 2000).

4   Figure 6.3 is from C.D. Keeling and T.P. Whorf, "Atmospheric CO_2 Concentrations (ppmv) Derived from In Situ Air Samples Collected at Mauna Loa Observatory, Hawaii," Scripps Institution of Oceanography, La Jolla, CA, August 16, 2000.

5   Paul N. Pearson and Martin R. Palmer, "Atmospheric Carbon Dioxide Concentrations over the Past 60 Million Years," *Nature*, vol. 406 (August 17, 2000): 695–9.

6   J.M. Barnola, D. Raynaud, C. Lorius and N.I. Barkov, "Historical CO_2 Record from the Vostok Ice Core," and J.R. Petit et al., "Historical Isotopic Temperature Record from the Vostok Ice Core," both in CDIAC, ORNL, *Trends: A Compendium of Data on Global Change* (Oak Ridge, TN, 1999).

7   Figure 6.4 is from James Hansen et al., Goddard Institute for Space Studies, "Global Temperature Anomalies in .01 °C," viewed on August 15, 2000 (www.giss.nasa.gov/data/update/gistemp).

8   M.E. Mann, R.S. Bradley and M.K. Hughes, "Northern Hemisphere Influences during the Past Millennium: Inferences, Uncertainties, and Limitations," *Geophysical Research Letters*, vol. 26, no. 6 (March 15, 1999): 759–62.

9   Thomas J. Crowley, "Causes of Climate Change over the Past 1000 Years," *Science*, vol. 289 (July 14, 2000): 270–77.

10   Hansen cited in Richard A. Kerr, "Globe's 'Missing Warming' Found in the Ocean," *Science*, vol. 287 (March 24, 2000): 2126–7.

11   Robert T. Watson, Marufu C. Zinyowera and Richard H. Moss (eds), *Climate Change 1995: Impacts, Adaptations and Mitigation of Climate Change*, Contribution of Working Group II to the Second Assessment Report of the Intergovernmental Panel on Climate Change (Cambridge, UK: Cambridge University Press, 1996).

12   D.A. Rotchrock, Y. Yu, and G.A. Maykut, "Thinning of the Arctic Sea-Ice Cover," *Geophysical Research Letters*, vol. 26, no. 23 (December 1, 1999): 1–5; Lars H. Smedsrud and Tore Furevik, "Towards an Ice-Free Arctic?" *Cicerone*, vol. 2 (February 2000): 1–7.

13   W. Krabill, W. Abdalati, E. Frederick, S. Manizade, C. Martin,

J. Sonntag, R. Swift, R. Thomas, W. Wright and J. Yungel, "Greenland Ice Sheet: High-Elevation Balance and Peripheral Thinning," *Science*, vol. 289 (July 21, 2000): 428–30.

14 Carsten Ruhlemann, Stefan Mulitza, Peter Müller, Gerold Wefer and Rainer Zahn, "Warming of the Tropical Atlantic Ocean and Slowdown of Thermohaline Circulation During the Last Glaciation," *Nature*, vol. 402 (December 2, 1999): 511–14.

15 Nebojsa Nakicenovic et al., *Special Report on Emission Scenarios*, Summary for Policymakers, Special Report of Working Group III of the Intergovernmental Panel on Climate Change (Geneva, Switzerland: May 2000).

16 Tom M.L. Wigley, *The Science of Climate Change: Global and US Perspectives* (Arlington, VA: Pew Center on Global Climate Change, June 29, 1999).

17 James Hansen et al., "Global Warming in the Twenty-First Century: An Alternative Scenario," *Proceedings of the National Academy of Sciences*, vol. 97 (June 16, 2000).

18 Hansen ct al., op. cit. (Note 17), 1.

19 Hadley Centre for Climate Prediction and Research, *Climate Change and its Impacts: Stabilisation of CO_2 in the Atmosphere* (London, UK: October 1999).

20 Michael Grubb with Christiaan Vrolijk and Duncan Brack, *The Kyoto Protocol: A Guide and Assessment* (London: Royal Institute for International Affairs, 1999).

21 J.T. Houghton et al. (eds), *Climate Change 1995: The Science of Climate Change*, Contribution of Working Group I to the Second Assessment Report of the Intergovernmental Panel on Climate Change (Cambridge, UK: Cambridge University Press, 1996).

22 Nebojsa Nakicenovic, Arnulf Grübler and Alan McDonald (eds), *Global Energy Perspectives* (Cambridge, UK: Cambridge University Press, 1998), 13.

23 Figure 6.5 is based on United Nations, *World Energy Supplies 1950–74* (New York, NY, 1976) and on BP Amoco, op. cit. (Note 3).

24 Nakicenovic, Grübler and McDonald, op. cit. (Note 22), 61.

25 Figure 6.6 is based on BP Amoco, op. cit. (Note 3), on Angus Maddison, *Monitoring the World Economy, 1820–1992* (Paris: Organisation for Economic Co-operation and Development, 1995),

and on International Monetary Fund (IMF), *World Economic Outlook* (Washington, DC, May 2000).

26 BP Amoco, op. cit. (Note 3), 33.

27 Ibid.

28 Ibid.

29 BP Amoco, op. cit. (Note 3), 9.

30 Ibid.

31 BP Amoco, op. cit. (Note 3), 25.

32 Ibid.

33 Ibid.

34 Ulrich Bartsch and Benito Müller, *Fossil Fuels in a Changing Climate* (Oxford, UK: Oxford University Press, 2000), 161–213.

35 Nakicenovic, Grübler, and McDonald, op. cit. (Note 22), 35–43.

36 Norman Myers and Jennifer Kent, *Perverse Subsidies: How Misused Tax Dollars Harm the Environment and Economy* (Washington, DC: Island Press, 2000).

37 Nakicenovic, Grübler and McDonald, op. cit. (Note 22), 35–43.

38 President's Committee of Advisors on Science and Technology (PCAST), *Federal Energy Research and Development for the Challenges of the Twenty-First Century* (Washington, DC, November 1997), 3-1–3-2.

39 Figure 6.7 is based on BP Amoco, op. cit. (Note 3), on Maddison, op. cit. (Note 25) and on IMF, op. cit. (Note 25).

40 Figure 6.8 is based on BP Amoco, op. cit. (Note 3), on Maddison, op. cit. (Note 25) and on IMF, op. cit. (Note 25).

41 Figure 6.9 is based on BP Amoco, op. cit. (Note 3), on Maddison, op. cit. (Note 25) and on IMF, op. cit. (Note 25).

42 Nakicenovic, Grübler and McDonald, op. cit. (Note 22), 35–43.

43 President's Committee of Advisors on Science and Technology (PCAST), *Powerful Partnerships: The Federal Role in International Cooperation on Energy Innovation* (Washington, DC, June 1999), chapters 4, 5.

44 PCAST, op. cit. (Note 43), chapter 4, 7–13.

45 PCAST, op. cit. (Note 43), chapter 4, 16–22.

46 Alex Taylor III, "Another Way to Beat High Gas Prices," *Fortune*, October 30, 2000, 58.

47 PCAST, op. cit. (Note 43), chapter 4, 23–31.

48 See Mark P. Mills and Peter Huber, "Dig More Coal: The PCs are Coming," *Forbes*, May 31, 1999, 70–72, and Joseph Romm with Arthur

Rosenfeld and Susan Herrmann, *The Internet Economy and Global Warming* (Arlington, VA: Center for Energy and Climate Solutions, December 1999), 5–8.

49 Romm, Rosenfeld and Herrmann, op. cit. (Note 48), 5–8.

50 Figure 6.10 is based on BP Amoco, op. cit. (Note 3), on Worldwatch database compiled from statistics from the International Atomic Energy Agency, on BTM Consult, *International Wind Energy Development: World Market Update 1999* (Ringkøbing, Denmark, March 2000), on Paul Maycock, "2000 World Cell/Module Production," *PV News*, March 2001, and on John Lund, "World Status of Geothermal Energy Use – Past and Potential," *Renewable Energy World*, July–August 2000, 122–31.

51 BTM Consult, op. cit. (Note 50).

52 Janet Ginsburg, "Green Power is Gaining Ground," *Business Week*, October 9, 2000, 44–5.

53 Ibid.

54 Paul Maycock, "The World PV Market 2000: Shifting from Subsidy to 'Fully Economic'?" *Renewable Energy World*, July–August 2000, 58–74.

55 Worldwatch estimate based on BP Amoco, op. cit. (Note 3), 37, on Lund, op. cit. (Note 50), 122–31, and on International Energy Agency (IEA), *Biomass Energy: Data, Analysis, and Trends* (Paris: IEA/OECD, 1998).

56 Shell International Petroleum Company, *The Evolution of the World's Energy System 1860–2060* (London, UK: Shell Centre, December 1995).

57 Shell Renewables, *Shell Renewables: Summary of Activities* (London: June 15, 2000).

58 Table 6.1 is based on Shell Renewables, op. cit. (Note 57), on Watson, Zinyowera and Moss, op. cit. (Note 11), 621–38, on Nakicenovic, Grübler and McDonald, op. cit. (Note 22), 35–43, on Paul Raskin et al., *Bending the Curve: Toward Global Sustainability* (Stockholm, Sweden: Stockholm Environment Institute, 1998), 49–56, on Bent Sorensen, *Long-Term Scenarios for Global Energy Demand and Supply* (Roskilde, Denmark: Roskilde University, January 1999), ii, 77–150.

59 Nakicenovic, Grübler and McDonald (eds), op. cit. (Note 22), 35–43.

60 Raskin et al., op. cit. (Note 58), 49–56.

61 Raskin et al., op. cit. (Note 58), 49–56.

62 Sorensen, op. cit. (Note 58), 77–150.

63 Ibid.

64 Sorensen, op. cit. (Note 58), 140–50.

65 Table 6.2 is from Curtis Moore and Jack Ihle, *Renewable Energy Policy Outside the United States*, Issue Brief No. 14 (Washington, DC: Renewable Energy Policy Project, October 1999).

66 BTM Consult, *Wind Force 10: A Blueprint to Achieve 10% of the World's Electricity from Wind Power by 2020* (London, UK: October 1999).

67 Figure 6.11 is based on Jesse H. Ausubel, "Can Technology Spare the Earth?" *American Scientist*, March–April 1996, 166–78.

68 "The Future of Fuel Cells," *Scientific American*, vol. 283, no. 7 (July 1999): 72–93.

69 Tom Koppel and Jay Reynolds, "A Fuel Cell Primer: The Promise and the Pitfalls," viewed on September 15, 2000, 26 (www.tomkoppel.com).

70 "DaimlerChrysler Offers First Commercial Fuel Cell Buses to Transit Agencies, Deliveries in 2002," *Hydrogen and Fuel Cell Letter*, May 2000, 1–2.

71 "New Fuel Cell Prototypes, Concepts on Display at Frankfurt, Tokyo Auto Shows," *Hydrogen and Fuel Cell Letter*, October 1999, 1–2.

72 "California Air Board Stays Course on ZEV Rule, First Cars Required by 2003," *Hydrogen and Fuel Cell Letter*, February 2001, 1–2.

73 California Fuel Cell Partnership, "California Fuel Cell Partnership Opens Headquarters Facility," press release (Sacramento, CA, November 1, 2000).

74 Bragi Arnason and Thorsteinn Sigfusson, "Iceland: A Future Hydrogen Economy," *International Journal of Hydrogen Energy*, vol. 25, no. 5 (May 2000): 389–94.

75 Seth Dunn, "The Hydrogen Experiment," *World Watch*, November/ December 2000, 14–25.

76 C.E.G. Padro and V. Putsche, *Survey of the Economics of Hydrogen Technologies* (Golden, CO: NREL, September 1999).

77 Pembina Institute for Appropriate Development, *Climate-Friendly Hydrogen Fuel: A Comparison of the Life-Cycle Greenhouse Gas Emissions for Selected Fuel Cell Vehicle Hydrogen Production Systems* (Drayton Valley, AB, Canada: Pembina Institute and David Suzuki Foundation, March 2000), 26.

78 Marc W. Jensen and Marc Ross, "The Ultimate Challenge: Developing an Infrastructure for Fuel Cell Vehicles," *Environment*, September 2000, 10–22.

79   John A. Turner, "A Realizable Renewable Energy Future," *Science*, vol. 285 (July 30, 1999): 687–9.

80   J. Ohi, *Blueprint for Hydrogen Fuel Infrastructure Development* (Golden, CO: NREL, January 2000), 1–3.

81   Bartsch and Müller, op. cit. (Note 34), 201.

82   Peter Schwartz, Peter Leyden and Joel Hyatt, *The Long Boom: A Vision for the Coming Age of Prosperity* (Reading, MA: Perseus Books, 1999), 171–86.

83   Mary Fagan, "Sheikh Yamani Predicts Price Crash as Age of Oil Ends," *Sunday Telegraph* (London), June 25, 2000.

84   Watson, Zinyowera and Moss, op. cit. (Note 11), 630.

85   Ibid.

86   "Oil-Rich Dubai, Starting Point of BMW LH2 10-Car World Tour, Considers Hydrogen," *Hydrogen and Fuel Cell Letter*, March 2001, 1–2.

87   "Emirates to Spend \$46 Billion on Ecology R&D over 10 Years," *Hydrogen and Fuel Cell Letter*, April 2001, 5.

88   Michael R. Bowlin, "Clean Energy: Preparing Today for Tomorrow's Challenges," Speech before the Cambridge Energy Research Associates Eighteenth Annual Executive Conference, *Globality and Energy: Strategies for the New Millennium* (Houston, TX, February 9, 1999).

Chapter 8

1   Described in Robert Weiner, "Financial Innovation in an Emerging Market: Petroleum Derivatives Trading in the 19th Century," George Washington School of Business and Public Management Working Paper 98–64, September 1998.

2   Gregory P. Nowell, *Mercantile States and the World Oil Cartel 1900-1939* (Ithaca, NY: Cornell University Press, 1994).

3   Note that the figures show US wellhead prices, there being no publicly available data series for internationally traded crude, given the lack of spot crude trading at the time.

4   Michael C. Lynch, "The Impact of Future Oil Price Paths on Oil Market Vulnerability to Supply Disruptions," *Papers and Proceedings of the Eighth Annual North American Conference*, International Association of Energy Economists, November 1986.

5 See, for example, "Oil Prices Expected to be Firm," *New York Times*, May 1, 1998.

6 *Petroleum Intelligence Weekly*, November 25, 1985, 3.

7 Many would disagree, given the bad track record for long-term oil price forecasting. However, in Michael Lynch, *Oil Prices to 2000: The Economics of the Oil Market* (London: Economist Intelligence Unit, 1989) my most likely scenario was for Arab Light prices to drop by 2000 to $13–18.5 (in 1998 dollars), the trend of which looks very accurate over the decade. This, however, varied so much from the conventional wisdom that it was referred to as "heretical" by the *Petroleum Economist*, September 1989, 270.

8 In the June 2000 issue of the US Department of Energy's *Monthly Energy Review* (which appears in July), for example, US oil deliveries (approximately equal to demand) for May were shown, but disaggregate OPEC oil production was available only for March, and OECD oil consumption and stock data only for January, although the delays are shorter for the larger countries.

9 I am speaking here particularly about Americans, because that is the debate that I am experienced with. Possibly other regions are more enlightened.

10 Coal protection became especially strong after the expansion of oil production in the Middle East following World War II.

Adams, F.G. and P. Miovic. "On Relative Fuel Efficiency and the Output Elasticity of Energy Consumption in Western Europe." *Journal of Industrial Economics* vol. 17 (1968).

Albritton, D.L. et al. *Summary for Policymakers: A Report of Working Group I of the Intergovernmental Panel on Climate Change.* Intergovernmental Panel on Climate Change (Port Chester, NY: Cambridge University Press, 2001).

American Wind Energy Association. "Wind Energy Fact Sheet: Comparative Costs of Wind and Other Fuels," 2000 (www.awea.org).

Arnason, Bragi and Thorsteinn Sigfusson. "Iceland: A Future Hydrogen Economy." *International Journal of Hydrogen Energy* vol. 25, no. 5 (May 2000): 389–94.

Arthur, W. Brian. "Competing Technologies, Increasing Returns, and Lock-In by Historical Events." *The Economic Journal* vol. 99 (March 1989).

Attanasi, E.D. and D.H. Root. "The Enigma of Oil and Gas Field Growth." *American Association of Petroleum Geologists* vol. 78, no. 3 (March 1994): Table 1.

Ausubel, Jesse H. "Can Technology Spare the Earth?" *American Scientist* March–April 1996, 166–78.

Ausubel, Jesse H. "Where is Energy Going?" *The Industrial Physicist* vol. 6, no. 1 (February 2000): 16–20.

Baer, Paul, John Harte, B. Haya, Antonio Herzog, John Holdren, Nathan Hultman, Daniel Kammen, Richard Norgaard and L. Raymond. "Equity and Greenhouse Gas Responsibility." *Science* vol. 289 (2000).

Barnola, J.M., D. Raynaud, C. Lorius and N.I. Barkov. "Historical CO_2 Record from the Vostok Ice Core," in Carbon Dioxide Information Analysis Center, Oak Ridge National Laboratory, *Trends: A Compendium of Data on Global Change* (Oak Ridge, TN: CDIAC, 1999).

Bartsch, Ulrich and Benito Müller. *Fossil Fuels in a Changing Climate* (Oxford, UK: Oxford University Press, 2000).

Bauen, A. and D. Hart. "Assessment of the Environmental Benefits of Transport and Stationary Fuel Cells." *Journal of Power Sources* vol. 86, no. 1–2 (2000).

Berndt, E.R. "Aggregate Energy, Efficiency, and Productivity Measurement." *Annual Review of Energy* vol. 3 (1978).

Blackburn, W. "Strategies for Commercializing Fuel Cell Vehicles in California." Fuel Cell Seminar 2000, Portland, Oregon (2000).

Boe, O., J. Flynn and E. Reiso. "On Near Wellbore Modeling and Real Time Reservoir Management." SPE 66369. Paper presented at SPE Reservoir Simulation Symposium, Houston, Texas, February 11–14, 2001.

Bowlin, Michael R. "Clean Energy: Preparing Today for Tomorrow's

Challenges." Speech before the Cambridge Energy Research Associates' Eighteenth Annual Executive Conference, "Globality and Energy: Strategies for the New Millennium," Houston, TX, February 9, 1999.

BP Amoco. *Statistical Review of World Energy* (London: Group Media & Publications, various years).

BTM Consult. *Wind Force 10: A Blueprint to Achieve 10% of the World's Electricity from Wind Power by 2020* (London, UK, October 1999).

BTM Consult. *International Wind Energy Development: World Market Update 1999* (Ringkøbing, Denmark, March 2000).

BTM Consult APS. "Wind Energy Statistics," 2000 (http://www.btm.dk/Statistics.htm).

California Fuel Cell Partnership. "California Fuel Cell Partnership Opens Headquarters Facility," press release (Sacramento, CA, November 1, 2000).

Campbell, C.J. and J.H. Laherrère. "The World's Oil Supply 1930–2050." Petroconsultants' Report, October 1995, 650.

Campbell, C.J. and J.H. Laherrère. "The End of Cheap Oil." *Scientific American* (March 1998): 80–85 (http://dieoff.com/page 140.htm).

Central Intelligence Agency. *International Energy Statistical Review* (1970–75).

Chapman, Jamie, Steven Wiese, Edgar DeMeo and Adam Serchuk. *Expanding Wind Power: Can Americans Afford It?* Renewable Energy Policy Project Report No. 6, Washington, DC, November 1998.

Cleveland, C.J. "Energy Quality and Energy Surplus in the Extraction of Fossil Fuels in the US." *Ecological Economics* vol. 6 (1992).

Cleveland, C.J., R. Kaufmann and D.I. Stern. "Aggregation and the Role of Energy in the Economy." *Ecological Economics* vol. 32 (2000).

Cleveland, C.J., R. Costanza, C.A.S. Hall and R. Kaufmann. "Energy and the US Economy: A Biophysical Perspective." *Science* vol. 255 (1984).

Collett, T.S. and K. Paull. "A Primer on Natural Gas Hydrates." Course 14 of the American Association of Petroleum Geologists' Annual Conference, New Orleans, April 16–19, 2000.

Cottrell, W.F. *Energy and Society* (New York, NY: McGraw-Hill, 1995).

Council of the European Union. *Council Directive 99/32/EC* (1999).

Cowan, Robin and David Kline. "The Implications of Potential 'Lock-In' in Markets for Renewable Energy." International Symposium on Energy and Environmental Management, Newport Beach, California, 1996.

Crosse, J. "Almost the Real Thing." *FT Automotive World*, May 1999, 40–43.

Crowley, Thomas J. "Causes of Climate Change Over the Past 1000 Years." *Science* vol. 289 (July 14, 2000): 270–77.

Department of Petroleum Engineering, The University of Tulsa Newsletter, October 2000.

Department of Trade and Industry. *UK Statistics 2000* (http://www2.dti.gov.uk/epa/intro.pdf).

Duke, Richard and Daniel Kammen. "The Economics of Energy Market

Transformation Initiatives." *The Energy Journal* vol. 20 (1999).

Duke, Richard, S. Graham, M. Hankins, A. Jacobson, D. Kammen, D. Khisa, D. Kithokoi, F. Ochieng, B. Osawa, S. Pulver and E. Walther. *Field Performance Evaluation of Amorphous Silicon (a-Si) Photovoltaic Systems in Kenya: Methods and Measurements in Support of a Sustainable Commercial Solar Energy Industry* (Washington, DC: The World Bank, 2000).

Dunn, Seth. "The Hydrogen Experiment." *World Watch* vol. 13, no. 6 (November/December 2000): 14–25.

Economides, M.H. "The State of R and D in the Petroleum Industry." *Journal of Petroleum Technology* vol. 47, no. 7 (July 1995).

Egusquiza, M.I. "3–D Seismic Surveys Improve the Cost-Benefit Ratio." SPE 69462. Paper presented at Society of Petroleum Engineers' Latin American and Caribbean Petroleum Engineering Conference, Buenos Aires, Argentina, March 25–28, 2001.

Energy Information Administration. *International Energy Outlook 2000* (Washington, DC: EIA, 2000).

Energy Price Statistics (Pennwell Publishing, various issues).

European Commission. *Energy for the Future: Renewable Sources of Energy* (Brussels: European Commission, 1997).

European Commission. *European Union Energy Outlook to 2020* (Brussels: European Commission, November 1999).

ExxonMobil. "Hoover Diana Platform in Gulf of Mexico." (www.exxonmobil.com).

Fagan, Mary. "Sheikh Yamani Predicts Price Crash as Age of Oil Ends." *Sunday Telegraph* (London), June 25, 2000.

Flavin, Christopher and Nicholas Lenssen. *Power Surge: Guide to the Coming Energy Revolution* (New York, NY: W.W. Norton and Co., 1994).

Forsdyke, I.N. "Flow Assurance in Multiphase Environments." SPE 37237. Paper presented at SPE International Symposium on Oilfield Chemistry, Houston, Texas, February 18–21, 1997.

Fowler, R.M. "World Conventional Hydrocarbon Resources: How Much Remains to be Discovered and Where Is It?" 16th World Petroleum Congress, Robertson Research International, 2000 (http://www.robresint.com/wpc).

Georgescu-Roegen, N. "Energetic Dogma, Energetic Economics, and Viable Technology," in J.R. Moroney (ed.) *Advances in the Economics of Energy and Resources* (Greenwich,CT: JAI Press, 1982).

Gever, J., R. Kaufmann, D. Skole and C. Vorosmarty. *Beyond Oil: The Threat to Food and Fuel in the Coming Decades* (Cambridge, MA: Ballinger, 1986).

Ginsburg, G.D. "Challenging the Presence of Natural Gas Hydrate in the Messoyakha Pool." *American Association of Petroleum Geologists* vol. 77, no. 9 (1993): 1624.

Ginsburg, Janet. "Green Power is Gaining Ground." *Business Week*, October 9, 2000, 44–5.

Gregoire Padró, C. and V. Putsche. "Survey of the Economics of Hydrogen Technologies." Report *NREL/TP-570–27079* (Golden, CO: National Renewable Energy Laboratory, 1999).

Grubb, Michael with Christiaan Vrolijk and Duncan Brack. *The Kyoto Protocol: A Guide and Assessment* (London: Royal Institute for International Affairs, 1999).

Grübler, A., N. Nakicenovic and D. Victor. "Dynamics of Energy Technologies and Global Change." *Energy Policy* vol. 27, no. 5 (1999).

Hadley Centre for Climate Prediction and Research. *Climate Change and Its Impacts: Stabilisation of CO_2 in the Atmosphere* (London, UK: October 1999).

Hall, C.A.S., C.J. Cleveland and R.K. Kaufmann. *Energy and Resource Quality: The Ecology of the Economic Process* (New York, NY: Wiley Interscience, 1986).

Hall, C.A.S., J.A. Stanford and F. Richard Hauer. "The Distribution and Abundance of Organisms as a Consequence of Energy Balances Along Multiple Environmental Gradients." *OIKOS* vol. 65 (1992).

Hansen, James et al. "Global Warming in the Twenty-First Century: An Alternative Scenario." *Proceedings of the National Academy of Sciences* vol. 97 (June 16, 2000).

Hansen, James et al. "Global Temperature Anomalies in .01 °C." Goddard Institute for Space Studies, viewed on August 15, 2000 (www.giss.nasa.gov/data/update/gistemp).

Hart, D. *Hydrogen Power: The Commercial Future of "the Ultimate Fuel"* (London, UK: Financial Times Energy Publishing, 1997).

Hart, D. and A. Bauen. *Fuel Cells: Clean Power, Clean Transport, Clean Future* (London: Financial Times Energy Publishing, 1997).

Hart, D., M.A. Leach, R. Fouquet, P.J. Pearson and A. Bauen. "Methanol Infrastructure: Will it Affect the Introduction of SPFC Vehicles?" *Journal of Power Sources* vol. 86, no. 1–2 (2000).

Hartley, L. *The Go-Between* (Harmondsworth, UK: Penguin, 1999).

Heuer, W. "The TES Project: A Joint Initiative for an Additional Fuel Infrastructure." *Journal of Power Sources* vol. 86, no. 1–2 (2000).

Hirsh, R. *Power Loss: The Origins of Deregulation and Restructuring in the American Utility System* (Cambridge, MA: MIT Press, 2000).

Höhlein, B., S. von Andrian, T. Grube and R. Menzer. "Critical Assessment of Power Trains with Fuel Cell Systems and Different Fuels." *Journal of Power Sources* vol. 86, no. 1–2 (2000).

Houghton, J.T., L.G. Meira Filho, B.A. Callender, N. Harris, A. Kattenberg and K. Maskell (eds) *Climate Change 1995: The Science of Climate Change.* Contribution of Working Group I to the Second Assessment Report of the Intergovernmental Panel on Climate Change (Cambridge, UK: Cambridge University Press, 1996).

Hovland, M. "Are there Commercial Deposits of Methane Hydrates in Ocean

Sediments?" Japanese National Oil Company "Methane Hydrates: Resources in the Near Future?" JNOC-TRC, Oct. 20–22, 1998.

Hubbert, M.K. "Energy Resources," in National Academy of Sciences, *Resources and Man* (San Francisco, CA: Freeman, 1969).

Hughes, T. *Networks of Power: Electrification in Western Society, 1870–1930* (Baltimore, MD: Johns Hopkins University Press, 1983).

Intergovernmental Panel on Climate Change Working Groups II and III. *Methodological and Technological Issues in Technology Transfer* (Cambridge, UK: Cambridge University Press, 2000).

International Energy Agency. *World Energy Outlook* (Paris: IEA/OECD, 1982).

International Energy Agency. *Oil, Gas and Coal Supply Outlook* (Paris: IEA/OECD, 1995).

International Energy Agency. *Oil Market Report* (Paris: IEA/OECD, 1995).

International Energy Agency. *Biomass Energy: Data, Analysis, and Trends* (Paris: IEA/OECD, 1998a).

International Energy Agency. *Natural Gas Information* (Paris: IEA/OECD, 1998b).

International Energy Agency. *World Energy Outlook* (Paris: IEA/OECD, 1998c).

International Energy Agency. CO_2 *Emissions from Fossil Fuel Combustion, 1971–1996* (Paris: IEA/OECD, 1999).

International Energy Agency. *Oil Market Report* (Paris: IEA/OECD, 2000).

International Institute for Applied Systems and Analysis/World Energy Council. *Global Energy Perspectives to 2050 and Beyond* (London: IIASA/WEC, 1995).

International Monetary Fund. *World Economic Outlook* (Washington, DC, May 2000).

Islas, Jorge. "Getting Round the Lock-In in Electricity Generating Systems: The Example of the Gas Turbine." *Research Policy* vol. 26 (1997).

Jack, I. "The E-Field (The Electric Oilfield)." SPE 68189. Paper presented at the Society of Petroleum Engineers' Middle East Oil Show, Bahrain, March 17–20, 2001.

Jensen, Marc W. and Marc Ross. "The Ultimate Challenge: Developing an Infrastructure for Fuel Cell Vehicles." *Environment* vol. 42, no. 7 (September 2000): 10–22.

Johansson, Thomas, Henry Kelly, Amulya Reddy and Robert Williams (eds) *Renewable Energy: Sources for Fuels and Electricity* (Washington, DC: Island Press, 1993).

Kamaruddin, S., A. Good and T. Gjerdingen. "Pushing the Envelope – Extending the Limits of Current Drilling Technology." SPE 64696. Paper presented at the Society of Petroleum Engineers' International Oil and Gas Conference and Exhibition, Beijing, China, November 7–10, 2000.

Kammen, Daniel. "Bringing Power to the People: Promoting Appropriate Energy Technologies in the Developing World." *Environment* vol. 41, no. 5 (1999).

Kaufmann, R.K. "A Biophysical Analysis of the Energy/Real GDP Ratio:

Implications for Substitution and Technical Change." *Ecological Economics* vol. 6 (1992).

Kaufmann, R.K. "The Relation Between Marginal Product and Price in US Energy Markets." *Energy Economics* vol. 16 (1994).

Keeling, C.D. and T.P. Whorf. "Atmospheric CO_2 Concentrations (ppmv) Derived From In Situ Air Samples Collected at Mauna Loa Observatory, Hawaii." Scripps Institution of Oceanography, La Jolla, CA, August 16, 2000.

Kempton, Willett and Steven Letendre. "Electric Vehicles as a New Power Source for Electric Utilities." *Transportation Research – D* vol. 2, no. 3 (1997).

Kerr, Richard A. "Globe's 'Missing Warming' Found in the Ocean." *Science* vol. 287 (March 24, 2000): 2126–7.

Kinzig, A. and D. Kammen. "National Trajectories of Carbon Emissions: Analysis of Proposals to Foster the Transition to Low-Carbon Economies." *Global Environmental Change* vol. 8, no. 3 (1998).

Koppel, Tom and Jay Reynolds. "A Fuel Cell Primer: The Promise and the Pitfalls," viewed on September 15, 2000 (www.tomkoppel.com).

Kordesch, K. and G. Simader. *Fuel Cells and Their Applications* (Weinheim, Germany: VCH Verlaggesellschaft, 1996).

Krabill, W., W. Abdalati, E. Frederick, S. Manizade, C. Martin, J. Sonntag, R. Swift, R. Thomas, W. Wright and J. Yungel. "Greenland Ice Sheet: High-Elevation Balance and Peripheral Thinning." *Science* vol. 289 (July 21, 2000): 428–30.

Laherrère, J.H. "Distributions de Type Fractal Parabolique dans la Nature." *Comptes Rendus de l'Académie des Sciences*, T.322, Series IIa, no. 7–4 (April 1996): 535–41 (http://www.oilcrisis.com/laherrere/fractal.htm).

Laherrère, J.H. "Assessing the Oil and Gas Future Production and the End of Cheap Oil?" *Canadian Society of Exploration Geophysicists*, Calgary (April 6, 1999) (http://dieoff.com/page179.htm).

Laherrère, J.H. "Reserve Growth: Technological Progress, or Bad Reporting and Bad Arithmetic?" *Geopolitics of Energy* vol. 22, no. 4 (April 1999): 7–16 (http://dieoff.com/page176.htm).

Laherrère, J.H. "Parabolic Fractal, Creaming Curve Improve Estimate of US Gulf Reserves." *Offshore Magazine* (May 1999): 113, 114, 177.

Laherrère, J.H. "Gas Hydrates," "Uncertain Resource Size Enigma" and "The SOFAR Channel: What and Why?" *Offshore Magazine* Part 1 (August 1999): 140–41 and 160–62.

Laherrère, J.H. "Data Shows Oceanic Methane Hydrate Resource Over-Estimated," *Offshore Magazine* Part 2 (September 1999): 156–8 (http://dieoff.com/page192.htm).

Laherrère, J.H. "Oceanic Hydrates: More Questions than Answers." *Energy Exploration and Exploitation*, Special Issue on Hydrates, 2000.

Laherrère, J.H. "Distribution of Field Sizes in a Petroleum System: Lognormal, Parabolic Fractal or Stretched Exponential?" *Marine and Petroleum Geology* vol. 17, no. 4 (April 2000): 539–46 (http://www.elsevier.nl/cgibin/cas/ tree/store/jmpg/cas_free/browse/browse.cgi?year=2000&volume=17&issue=4).

Laherrère, J.H. "Learn Strengths, Weaknesses to Understand Hubbert Curve." *Oil and Gas Journal* (April 17, 2000) (http://dieoff.com/page191.htm).

Laherrère, J.H. "Is the USGS 2000 Assessment Reliable?" Cyberconference by the World Energy Council on Strategic Options, May 19, 2000 (http://www.energyresource2000.com).

Laherrère, J.H., A. Perrodon and G. Demaison. "Undiscovered Petroleum Potential." Petroconsultants' Report, 1994, 383.

Larminie, J. and Dicks, A. *Fuel Cell Systems Explained* (Chichester: Wiley, 2000).

Lipman, Timothy. *Zero-Emission Vehicle Scenario Cost Analysis Using a Fuzzy Set-Based Framework* (Davis, CA: Institute of Transportation Studies, December 1999).

Lipman, Timothy and Kammen, D.M. "Renewable Energy: Now a Realistic Challenge to Oil." *The Future of Oil as a Source of Energy* (Abu Dhabi: The Emirates Center for Strategic Studies and Research, 2002).

Lovins, Amory. *Soft Energy Paths: Toward a Durable Peace* (Cambridge, MA: Ballinger Press, 1977).

Lund, John. "World Status of Geothermal Energy Use – Past and Potential." *Renewable Energy World*, vol. 3, no. 4 (July–August 2000): 122–31.

Lynch, Michael C. "The Impact of Future Oil Price Paths on Oil Market Vulnerability to Supply Disruptions." *Papers and Proceedings of the Eighth Annual North American Conference*, International Association of Energy Economists, November 1986.

Lynch, Michael C. *Oil Prices to 2000: The Economics of the Oil Market* (London: Economist Intelligence Unit, 1989).

Mabro, Robert. "The World's Oil Supply 1930–2050." A Review Article. *Journal of Energy Literature*, vol. II, no. 1 (1996): 32.

Maddison, Angus. *Monitoring the World Economy, 1820–1992* (Paris: OECD, 1995).

Mann, M.E., R.S. Bradley and M.K. Hughes, "Northern Hemisphere Influences During the Past Millennium: Inferences, Uncertainties, and Limitations." *Geophysical Research Letters* vol. 26, no. 6 (March 15, 1999): 759–62.

Margolis, Richard and Daniel Kammen. "Underinvestment: The Energy Technology and R&D Policy Challenge." *Science* vol. 285 (1999).

Marland, G., T.A. Boden and R.J. Andres, "Global, Regional, and National CO_2 Estimates from Fossil Fuel Burning, Cement Production, and Gas Flaring: 1751–1997" (revised August 22, 2000), Carbon Dioxide Information Analysis Center, Oak Ridge National Laboratory, Oak Ridge, TN, August 22, 2000.

Maycock, Paul. "The World PV Market 2000: Shifting from Subsidy to 'Fully Economic'?" *Renewable Energy World* vol. 3, no. 4 (July–August 2000): 58–74.

Maycock, Paul. "2000 World Cell/Module Production." *PV News*, March 2001.

Mills, Mark P. and Peter Huber. "Dig More Coal: The PCs are Coming." *Forbes*, May 31, 1999, 70–72.

Mjaaland, S., A. Wulff, E. Causse and F. Nyhavn. "Integrating Seismic Monitoring and Intelligent Wells." SPE 62878. Paper presented at the Society of Petroleum Engineers' Annual Technical Conference and Exhibition, Dallas, Texas, October 1–4, 2000.

Moore, Curtis and Jack Ihle, "Renewable Energy Policy Outside the United States," Issue Brief No. 14 (Washington, DC: Renewable Energy Policy Project, October 1999).

Moss, S.A. "Potential Carbon Emission Savings from Energy Efficiency in Commercial Buildings." Report *IP3/96* (Watford: Buildings Research Establishment, 1996).

Myers, Norman and Jennifer Kent. *Perverse Subsidies: How Misused Tax Dollars Harm the Environment and Economy* (Washington, DC: Island Press, 2000).

Nakicenovic, Nebojsa, Arnulf Grübler and Alan McDonald (eds) *Global Energy Perspectives* (Cambridge, UK: Cambridge University Press, 1998).

Nakicenovic, Nebojsa, Ogunlade Davidson, Gerald Davis, Arnulf Grübler, Tom Kram, Emilio Lebre La Rovere, Bert Metz, Tsuneyuki Morita, William Pepper, Hugh Pitcher, Alexei Sankovski, Priyadarshi Shukla, Robert Swart, Robert Watson and Zhou Dhadi. *Special Report on Emission Scenarios*. Summary for Policymakers, Special Report of Working Group III of the Intergovernmental Panel on Climate Change (Geneva, Switzerland: May 2000).

Nowell, Gregory P. *Mercantile States and the World Oil Cartel 1900–1939* (Ithaca, NY: Cornell University Press, 1994).

Nurmi, R. "Horizontal Highlights," in *Middle East Evaluation Review* (Schlumberger Publications, 1995).

Nuveen Investments. *Nuveen Defined Portfolios Product Guide*, company literature (June 2000).

Odum, H.T. "Trophic Structure and Productivity of Silver Springs, Florida." *Ecological Monographs* vol. 27 (1957).

Odum, H.T. *Environment, Power and Society* (New York, NY: Wiley-Interscience, 1971).

Ogden, Joan. "Developing an Infrastructure for Hydrogen Vehicles: A Southern California Case Study." *International Journal of Hydrogen Energy* vol. 24, no. 8 (1999).

Ogden, Joan and Robert Williams. *Solar Hydrogen: Moving Beyond Fossil Fuels* (Washington, DC: World Resources Institute, 1989).

Ogden, Joan, Eric Larson and Mark Delucchi. *A Technical and Economic Assessment of Renewable Transportation Fuels and Technologies* (Washington, DC: Office of Technology Assessment, May 1994).

Ohi, J. *Blueprint for Hydrogen Fuel Infrastructure Development* (Golden, CO: National Renewable Energy Laboratory, January 2000), 1–3.

Ottinger, Richard, David Wooley, David Hodas, Nicholas Robinson and Susan Babb. *Environmental Costs of Electricity* (New York, NY: Oceana Publications, Inc., 1991).

Padro, C.E.G. and V. Putsche. *Survey of the Economics of Hydrogen Technologies* (Golden, CO: National Renewable Energy Laboratory, September 1999).

Papathanasiou, D. and D. Anderson. *Uncertainties in Responding to Climate Change: On the Economic Value of Technology Policies for Reducing Costs and Creating Options* (London: Imperial College Centre for Energy Policy and Technology, 2000).

Pasicznyk, A. "Evolution Toward Simpler, Less Risky Multilateral Wells." SPE 67825. Paper presented at Society of Petroleum Engineers/International Association of Drilling Companies Drilling Conference, Amsterdam, The Netherlands, February 27–March 1, 2001.

Patterson, W. *Transforming Electricity* (London: Earthscan, 1999).

Pearce, F. "Back to the Days of Deadly Smogs." *Nature* no. 1850 (Dec. 5, 1992).

Pearson, Paul N. and Martin R. Palmer. "Atmospheric Carbon Dioxide Concentrations Over the Past 60 Million Years." *Nature* no. 406 (August 17, 2000): 695–9.

Pembina Institute for Appropriate Development. *Climate-Friendly Hydrogen Fuel: A Comparison of the Life-Cycle Greenhouse Gas Emissions for Selected Fuel Cell Vehicle Hydrogen Production Systems* (Drayton Valley, AB, Canada: Pembina Institute and David Suzuki Foundation, March 2000), 26.

Perrodon, A., J.H. Laherrère and C.J. Campbell. "The World's Non-Conventional Oil and Gas." *Petroleum Economist* March 1998, 113 (http://textor.com/cms/dPEWNC.html).

Petit, J.R. et al. "Historical Isotopic Temperature Record from the Vostok Ice Core," in Carbon Dioxide Information Analysis Center, Oak Ridge National Laboratory, *Trends: A Compendium of Data on Global Change* (Oak Ridge, TN: CDIAC, 1999).

Petroleum Intelligence Weekly, vol. XXXVIII, no. 49, December 1, 1999.

Post, Mac. "Global Carbon Cycle (1992–1997)." Carbon Dioxide Information Analysis Center, Oak Ridge National Laboratory, viewed on September 7, 2000 (cdiac.esd.ornl.gov/pns/graphics/globcarb.gif).

President's Committee of Advisors on Science and Technology. *Federal Energy Research and Development for the Challenges of the Twenty-First Century.* Report of the PCAST Energy Research and Development Panel (Washington, DC: November, 1997).

President's Committee of Advisors on Science and Technology. *Powerful Partnerships: The Federal Role in International Cooperation on Energy Innovation* (Washington, DC: June 1999).

Rambach, G.D. "An Examination of Isolated, Stationary, Hydrogen Power

Systems Supplied by Renewables." 9th Canadian Hydrogen Conference, Vancouver, Canada, 1999.

Raskin, Paul et al. *Bending the Curve: Toward Global Sustainability* (Stockholm, Sweden: Stockholm Environment Institute, 1998).

Rennie, J. "13 Vehicles That Went Nowhere." *Scientific American* vol. 277, no. 4 (1997).

Reuters News Service. "Fuel Cells and New Energies Come of Age Amid Fuel Crisis," September 11, 2000 (http://www.reuters.com).

Romm, Joseph with Arthur Rosenfeld and Susan Herrmann, *The Internet Economy and Global Warming* (Arlington, VA: Center for Energy and Climate Solutions, December 1999).

Rotchrock, D.A., Y. Yu and G.A. Maykut. "Thinning of the Arctic Sea-Ice Cover." *Geophysical Research Letters* vol. 26, no. 23 (December 1, 1999): 1–5.

Ruhlemann, Carsten, Stefan Mulitza, Peter Müller, Gerold Wefer and Rainer Zahn. "Warming of the Tropical Atlantic Ocean and Slowdown of Thermohaline Circulation During the Last Glaciation." *Nature* no. 402 (December 2, 1999): 511–14.

Schumacher, E.F. *Small is Beautiful* (London: Vintage, 1993).

Schwartz, Peter, Peter Leyden and Joel Hyatt. *The Long Boom: A Vision for the Coming Age of Prosperity* (Reading, MA: Perseus Books, 1999).

Shell International Petroleum Company. *The Evolution of the World's Energy System 1860–2060* (London, UK: Shell Centre, December 1995).

Shell International Petroleum Company. *Global Scenarios: 1998–2020*, Summary Brochure (London, UK: Shell Centre, 2000).

Shell Renewables. *Shell Renewables: Summary of Activities* (London, UK, June 15, 2000).

Simmons, M.R. "Fighting Rising Demand and Rising Decline Curves: Can the Challenge Be Met?" Society of Petroleum Engineers' Asia Pacific Oil and Gas Conference, Yokohama, Japan, April 25, 2000 (http://www.simmonsco-intl.com/web/downloads/spe.pdf).

Smedsrud, Lars H. and Tore Furevik. "Towards an Ice-Free Arctic?" *Cicerone* vol. 2 (February 2000): 1–7.

Smil, V. *General Energetics* (New York, NY: Wiley-Interscience, 1991).

Sorensen, Bent. *Long-Term Scenarios for Global Energy Demand and Supply* (Roskilde, Denmark: Roskilde University, January 1999).

Spence, M. "The Learning Curve and Competition." *Bell Journal of Economics* vol. 12 (1981).

Stern, D.I. "Energy Use and Economic Growth in the USA: A Multivariate Approach." *Energy Economics* vol. 15 (1993).

Sumrow, M.H. and M.J. Economides. "Pushing the Boundaries of Coiled Tubing Applications." SPE 68480. Paper presented at the Society of Petroleum Engineers/International Coil Tubing Association Coiled Tubing Roundtable, Houston, Texas, March 7–8, 2001.

Swezey, B.G. and Yih-huei Wan. *The True Cost of Renewables: An Analytic Response to the Coal Industry's Attack on Renewable Energy* (Golden, CO: US Department of Energy, 1996).

Taylor III, Alex. "Another Way to Beat High Gas Prices." *Fortune*, October 30, 2000, 58.

Tempest, P. *World Petroleum at the Crossroads* (London, UK: PTA Greenwich, 1999).

Ter-Gazarian, A. *Energy Storage for Power Systems* (London, UK: Peter Pelegrinus Ltd, on behalf of Institution of Electrical Engineers, 1994).

Thomas, C., B. James, F. Lomax and I. Kuhn. "Integrated Analysis of Hydrogen Passenger Vehicle Transportation Pathways." Report *DE-AC36–83CH10093*, Golden, CO, National Renewable Energy Laboratory, 1998.

Turner, John A. "A Realizable Renewable Energy Future." *Science* vol. 285 (July 30, 1999): 687–9.

United Nations, *World Energy Supplies 1950–74* (New York, NY: 1976).

US Department of Energy. *Renewable Energy Technology Characterizations* (Washington, DC: US DOE Office of Utility Technologies and Electricity Policy Research Institute, 1997).

US Department of Energy. *Annual Energy Outlook 2000* (Washington, DC: US DOE Energy Information Administration, 2000).

US Department of Energy. *Annual Energy Review* (various issues).

US Federal Register. "National Ambient Air Quality Standards for Particulate Matter: Proposed Decision." vol. 61 (1996).

US Geological Survey 2000. "World Petroleum Assessment 2000 – Description and Results." 16th World Petroleum Congress, summary: (http://energy.cr.usgs.gov:8080/energy/WorldEnergy/weppdf/sumworld.xls).

Wang, J.H., Chiang, W.L., and Shu, J.P.H. "The Prospects: Fuel Cell Motorcycle in Taiwan." *Journal of Power Sources* vol. 86, no. 1–2 (2000).

Watanabe, Chihiro. "Industrial Dynamism and the Creation of a 'Virtuous Cycle' between R&D, Market Growth and Price Reduction: The Case of Photovoltaic Power Generation (PV) Development in Japan," in Clas-Otto Wene, A. Voss and T. Fried (eds) *Proceedings of the IEA International Workshop on Experience Curves for Policy Making: The Case of Energy Technologies*, Stuttgart, Germany, May 10–11, 1999 (Stuttgart: 2000).

Watson, Robert T., Marufu C. Zinyowera and Richard H. Moss (eds) *Climate Change 1995: Impacts, Adaptations and Mitigation of Climate Change*, Contribution of Working Group II to the Second Assessment Report of the Intergovernmental Panel on Climate Change (Cambridge, UK: Cambridge University Press, 1996).

Weiner, Robert. "Financial Innovation in an Emerging Market: Petroleum Derivatives Trading in the 19th Century." George Washington School of Business and Public Management Working Paper 98–64, September 1998.

Wigley, Tom M.L. *The Science of Climate Change: Global and US Perspectives* (Arlington, VA: Pew Center on Global Climate Change, June 29, 1999).

Wood, J. "Long Term World Oil Supply (A Resource Base/Production Path Analysis)," 2000 (http://www.eia.doe.gov/pub/oil_gas/petroleum/presentations/2000/long_term_supply/LongTermOilSupplyPresentation.ppt).

World Bank. *Global Commodity Markets*, vol. 7, no. 2 (April 2000).

World Bank. "Market Outlook for Major Primary Commodities." *Global Commodity Markets*, vol. 2, no. 1 (January 1994).

World Bank. *World Bank Indicators* (Washington, DC: World Bank, 1997).

World Energy Council. "Energy for Tomorrow's World – Acting Now" 2000, 80–81.

World Gas Intelligence. *World Gas Intelligence*, vol. 7, no. 20 (October 11, 1996).

Youngquist, Walter. *Geodestinies – The Inevitable Control of Earth Resources Over Nations and Individuals* (Portland, OR: National Book Company, 1997).

Youngquist, Walter. "Shale Oil: The Elusive Energy." *Hubbert Center Newsletter*, April 1998 (http://hubbert.mines.edu).